HOW TO PASS THE
RAE

SECOND EDITION

Clive Smith G4FZH

Edited by
George Benbow G3HB

RSGB

Radio Society of Great Britain

Published by the Radio Society of Great Britain, Cranborne Road, Potters Bar, Herts EN6 3JE.

Second edition 1989

ISBN 0 900612 86 X

Cover Design by Linda Penny and Neil Jackson, Radio Society of Great Britain.
Typography by Neil Jackson, Radio Society of Great Britain.
Printed by Bell & Bain Ltd., Glasgow.

Contents

Preface   4

1. What is a multiple-choice examination?   5

2. Tackling the multiple-choice RAE   7

3. Mathematics for the RAE   9

4. Preparing for the RAE   16

5. Practice multiple-choice examinations   19

Answers to questions   83

Appendix   84

Preface

How to pass the RAE is published by the Radio Society of Great Britain as a guide to would-be amateurs who intend to take the Radio Amateurs' Examination.

The RAE is a multiple choice examination and so chapter 1 is an explanation of this type of examination, its implications and how it compares with the earlier written examination.

Tackling the multiple choice examination is then discussed with a recommended approach to studying for it.

Mathematical questions in the RAE often present difficulties to the mature student and hence a revision course to the level of RAE mathematics is included in chapter 3.

The student is then taken through all the stages of preparing for the examination, from finding a college course, to the preliminary formalities in the examination room and how to fill in the answer sheet.

The sample examination papers in this book have been revised to meet the RAE syllabus for 1989-91, in particular the introduction of electromagnetic compatibility. This onerous task was carried out by Dr C. V. Smith, G4FZH, assisted by Helen Smith, G4KNQ.

The questions are in the RAE format and are closely representative of the scope of the examination in context and difficulty, but not exactly so as they have not been pre-tested and validated as those set by the City and Guilds of London Institute.

The permission of the City and Guilds of London Institute to reproduce the answer sheet is gratefully acknowledged.

George Benbow, G3HB
Editor

CHAPTER 1

What is a multiple choice examination?

For many years, educationalists, parents and students have argued against the written examination, how unfair it is, and so on. There is little point in repeating these arguments here.

An alternative to the written examination is "objective testing". An objective test can be defined as a series of questions, each of which has only one correct answer. There are several types of objective question, one of which is the four option multiple choice question.

Multiple choice question examinations came into use about 1968 and are now common at all levels of school and technical examinations.

The Radio Amateurs' Examination, the RAE, was originated in 1947 as a written examination conducted by the City and Guilds of London Institute on behalf of the amateur radio licensing authority. The RAE became an objective test using the four option multiple choice type of question in 1979.

The question arises: what does a multiple choice question look like? An easy way of answering is to consider a simple question as might be set in an RAE.

The correct Q-code to indicate that an operator wishes the sending station to send more slowly is:

   a. QSK
   b. QSY
   c. QRS
   d. QSS

Each of these Q codes has some meaning, at least two, QSY and QRS are in frequent use by the radio amateur. Another of them, QSS, could be guessed as "send slowly" but it is not the right answer. In fact, to the student who has prepared for the examination and who has bothered to learn the Q codes properly, or better still, who has listened as a shortwave listener (swl) for some time, there is no problem. To him, the answer is QRS and he will underline "c" on his answer sheet and gain a mark towards his RAE pass slip. It is a valid question, based on the syllabus, which sorts out the "guesser" from the "knower"

The multiple choice question is known as the "stem" and the four answer are the "options". The correct answer is called the "key" and the three incorrect ones known as "distractors".

The multiple choice examination therefore consists of a number of such questions having four possible answers. The candidate is required to indicate which he thinks is the correct one by filling one of four small blocks on the answer sheet, labelled a, b, c, and d with an HB pencil.

The completed paper is marked automatically by a computerised reading machine, which senses the pencil marks on the answer sheet.

Like all changes, it was not and is still not universally accepted as a good idea by the amateur fraternity. However, perhaps this non-acceptance was aided by the appearance of less than useful questions in the early examinations.

The following observations are intended to give some idea of the trouble the setters of multiple choice questions and papers have taken to ensure a good result. They should remove some of the common misconceptions about the present RAE.

All likely questions, after a preliminary check, are pre-tested by about 300 specially invited "guinea-pig" students at colleges throughout the country. This is followed by statistical analysis of their answers to each question.

One of the various aspects of a question that can be examined is how well the question discriminates between levels of ability and the likelihood that an able candidate will get question right or wrong. In simple terms, if a question is answered wrongly by 90 per cent of all those answering, it can be classified as "very hard". If 90 per cent get it right, it is "too easy".

Sometimes a question can be devised, with its answer and three reasonable distractors, that "catches out" the clever candidates. This means that an above-average number of candidates with an overall high mark get that particular item wrong. Again, this is not a question likely to produce a good paper and it would be left out of future papers. By keeping a check on these "hard/easy" performances of many questions of each type, a bank of many hundreds of questions is built up. Thus, by random choice of questions from the bank, a paper of almost uniform standard can be produced for each examination.

An additional bonus that arises from multiple choice is the removal of the slow, costly and difficult to understand marking procedure. With a written examination, what one examiner may regard as a good answer, another may not rate highly. He could have toothache and just feel tetchy when it comes to a grade boundary mark! What if that boundary is your pass/fail decision? Long meetings by the examiners to adjudicate on these decisions are bad enough even if only a few candidates are involved. For large numbers the expense and time become major problems.

To deal now with the old chestnut that you can get the right answers to pass or even obtain distinction level by guessing. There are excellent statistical methods of showing that for a 35 question paper with four possibles for each question the chance of picking a winner by guesswork is millions to one against! The average horse race punter may hope to bring off a win treble in three races with four horses in each race. To suggest that he always can do so is to invite cynical laughter from any group of gamblers! Try asking them to come up with a win 24 out of 35 races and you are liable too find yourself the object of great amusement. No, you cannot guess your way through these papers and pass.

The marking procedure, although automatic, is complex. After the first 300 answer sheets have been marked, a preliminary analysis of the results is made to check that the results for each question are as expected from the pre-test. If any question shows an unexpected result, it is disregarded. The answer sheets are finally marked after this check has been carried out. A sample of the automatically marked sheets is also checked manually.

The important aspect of what is being examined is knowledge and application of knowledge, rather than the testing of ability to express knowledge by coherent use of language and diagrams. The multiple choice format is the best way of producing a reasonable check of a student's ability. For the purposes of deciding if someone can be entrusted with privilege of operating on the amateur bands, it is an effective form of examination.

To summarise, of all the advantages claimed for the multiple choice examination, those likely to be of most concern to the RAE candidates are:

1. Because there are a large number of questions, all areas of the syllabus can be covered. A written paper can only cover a limited number of topics, typically eight in a three hour examination.

2. As the correct answer to each question is predetermined, marking is purely objective. There is no question of the marker having to decide how much credit to give for a part correct answer. The mood of the examiner at the time of marking or the difference between two examiners are of no significance at all in this context.

3. There is no requirement on the candidates to produce a well composed and legibly written answer of 300-400 words in 15 minutes or so, possibly with a couple of diagrams. This may be easy for some, but it is difficult for many students.

CHAPTER 2

Tackling the multiple choice RAE

This type of paper takes much less time than the old long answer questions. On average you have about 90 seconds to read the question, decide which is the correct answer, mark the answer sheet and go on to the next item! This means a special technique is needed if you are to make the best use of what is stored in your memory. You are recommended to proceed in the following way.

Learning and revising

Proceed in a logical fashion using the *RAE Manual* and notes and hand-outs given by the lecturer, or conversations with fellow enthusiasts to help you to gain as clear a grasp as you can of what is in the syllabus. Do not bother too much about the examination at this stage: concentrate on learning and, as far as possible, understanding the work.

Revision should be in two ways. As you go along make sure you check your day's or week's work. Try to retain what you have learned by revising previous work throughout the course. As the examination comes closer you can start to measure the success of your efforts by attempting some multiple choice questions from CæG and RSGB publications. In other words, do a "mock exam" at home, not in the rather forbidding environment of the examination room. Note that some of the books of questions published are rather inadequate, untested and unstandardised and sometimes may do more harm than good.

With the ready availability of sophisticated equipment merely by paying for it, and with the influx of enthusiasts who have not served an swl apprenticeship, it is possible to get on the air with little idea of how to operate properly. If you have learned what is in the syllabus, talked to experienced enthusiasts, and have engaged in some listening you are less likely to make an avoidable blunder. You will certainly enjoy the hobby more than those who "cram for exam" as far as the RAE is concerned.

The actual examination papers

Probably the most important single piece of advice is to read the questions. After all examinations, one meets candidates who have mis-read questions and have thrown away marks as a result. Remember, you will be under tension. You will probably have done some last minute cramming and up will pop a question that is the

subject of that work or at least pretty close to it. Before you can say "underline it " you will have fallen for a cleverly set distractor and away will go another mark.

The second piece of advice is not to spend time fretting over a tricky question. If you do not come up with an answer quickly, leave that question and go on to consider the next. This is most important. A consequence of this approach is that you will probably answer half of the items in you first pass through the paper. You can than start at the beginning and retrace your steps from first to last question. You can hope to answer perhaps half of the remaining questions on this run. Return and try to see what is required for those remaining. In the last couple of minutes you should fill in an answer for all remaining questions. There are no penalty marks for wrong answers and to leave and unanswered question, even if you have to guess, is not a good idea. The initial remarks about guessing apply to doing this for all questions. In any event, if you have worked at your RAE, few are likely to be blind guesses so the odds on a correct answer are better than you would expect for such a guess.

The third piece of advice is really a consequence of the recommended approach involving successive runs through the questions. The question booklet and the answer sheet are discrete items. It is very easy to get out of order between them. That is you underline 7c for answer to question 8. Once you have done this it tends to run on for several questions until, with horror, you notice what has happened. This is than followed by panic stricken rubbing out and than a distracted period of poor concentration as you try to get back to smooth work. Remember, time will be marching steadily on while this is happening. To avoid this experience, it is a good idea to use a spare pencil. If you are right handed, keep your left index finger very firmly pointing question 7... or what have you. The spare pencil than underlines question 7's space on the answer sheet. When you have marked question 7, your finger then moves to the next question you are going to tackle and your spare pencil underlines that question's space on the answer sheet. Ergo, no panic and you can hop from question to question with impunity.

To summarise, the flow diagram for the best possible use of your knowledge and efforts looks something like this:

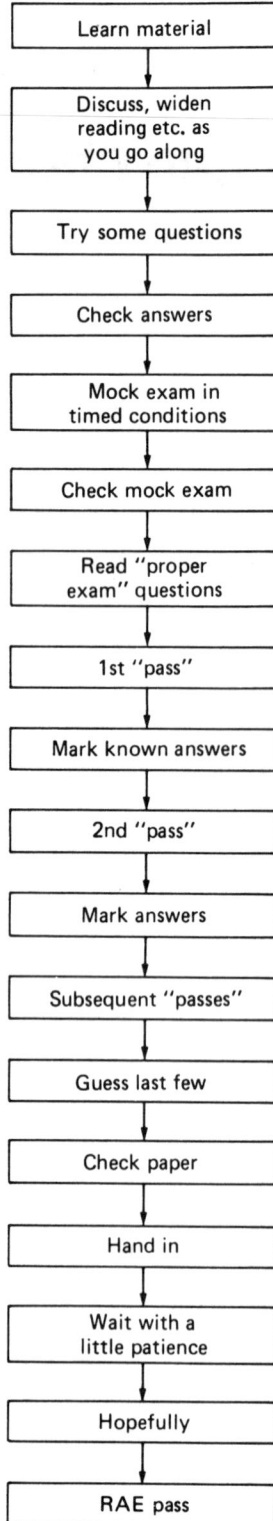

```
            ┌─────────────────────────┐
            │      Learn material      │
            └─────────────────────────┘
                        │
                        ▼
            ┌─────────────────────────┐
            │      Discuss, widen      │
            │      reading etc. as     │
            │      you go along        │
            └─────────────────────────┘
                        │
                        ▼
            ┌─────────────────────────┐
            │    Try some questions    │
            └─────────────────────────┘
                        │
                        ▼
            ┌─────────────────────────┐
            │       Check answers      │
            └─────────────────────────┘
                        │
                        ▼
            ┌─────────────────────────┐
            │       Mock exam in       │
            │      timed conditions    │
            └─────────────────────────┘
                        │
                        ▼
            ┌─────────────────────────┐
            │      Check mock exam     │
            └─────────────────────────┘
                        │
                        ▼
            ┌─────────────────────────┐
            │      Read "proper        │
            │      exam" questions     │
            └─────────────────────────┘
                        │
                        ▼
            ┌─────────────────────────┐
            │        1st "pass"        │
            └─────────────────────────┘
                        │
                        ▼
            ┌─────────────────────────┐
            │    Mark known answers    │
            └─────────────────────────┘
                        │
                        ▼
            ┌─────────────────────────┐
            │        2nd "pass"        │
            └─────────────────────────┘
                        │
                        ▼
            ┌─────────────────────────┐
            │       Mark answers       │
            └─────────────────────────┘
                        │
                        ▼
            ┌─────────────────────────┐
            │    Subsequent "passes"   │
            └─────────────────────────┘
                        │
                        ▼
            ┌─────────────────────────┐
            │      Guess last few      │
            └─────────────────────────┘
                        │
                        ▼
            ┌─────────────────────────┐
            │        Check paper       │
            └─────────────────────────┘
                        │
                        ▼
            ┌─────────────────────────┐
            │         Hand in          │
            └─────────────────────────┘
                        │
                        ▼
            ┌─────────────────────────┐
            │       Wait with a        │
            │      little patience     │
            └─────────────────────────┘
                        │
                        ▼
            ┌─────────────────────────┐
            │        Hopefully         │
            └─────────────────────────┘
                        │
                        ▼
            ┌─────────────────────────┐
            │         RAE pass         │
            └─────────────────────────┘
```

CHAPTER 3

Mathematics for the RAE

The basic mathematical processes are: addition, subtraction, multiplication, and division. As long as only "whole" numbers are involved, such sums are simple.

However, very often we must consider quantities which are less than one (unity), for instance, $\frac{1}{2}$, $\frac{1}{3}$, $\frac{1}{8}$ etc. Here $\frac{1}{8}$ means one-eighth part of the whole and so on.

$\frac{1}{8}$ is called a *vulgar fraction* and has two parts: the "8" (the bottom part) is called the *denominator* and the "1" (the top part) is the *numerator*. The magnitude of a fraction is not changed if we multiply top and bottom by the same number, ie

$$\frac{3}{16} \times \frac{4}{4}$$

As the "4" is on the top and the bottom we can "cancel" it thus:

$$\frac{3}{16} \times \frac{4}{4} = \frac{3}{16}$$

A fraction should always be cancelled down to its simplest form:

$$\frac{12}{16} = \frac{3 \times 4}{4 \times 4} = \frac{3}{4}$$

Here top and bottom have been divided by 4.

Fractions can be
(a) *Multiplied*

$$\frac{1}{2} \times \frac{3}{4} \times \frac{5}{8} = \frac{1 \times 3 \times 5}{2 \times 4 \times 8} = \frac{15}{64}$$

(b) *Divided*

$$\frac{3}{4} \div \frac{1}{2}$$

Dividing by $\frac{1}{2}$ is the same as multiplying by $\frac{2}{1}$, ie

$$\frac{3}{4} \div \frac{1}{2} = \frac{3}{4} \times \frac{2}{1} = \frac{6}{4} = 1\frac{2}{4} = 1\frac{1}{2}$$

In other words, dividing by a fraction is the same as multiplying by that fraction "upside down". Another example is:

$$\frac{7}{8} \div \frac{3}{4} = \frac{7}{8} \times \frac{4}{3} = \frac{7}{2} \times \frac{1}{3} = \frac{7}{6} = 1\frac{1}{6}$$

Here we divide top and bottom by 4.

(c) *Added*

$$\frac{2}{3} + \frac{2}{3} = \frac{2 + 2}{3} = \frac{4}{3}$$

If the denominators are different, we must make them the same, ie "bring them to a common denominator" normally the lowest common denominator is used. For example

$$\frac{2}{3} + \frac{5}{6} = \frac{4}{6} + \frac{5}{6} = \frac{9}{6} = \frac{3}{2} = 1\frac{1}{2}$$

Here we have multiplied top and bottom of $\frac{2}{3}$ by 2, making it $\frac{4}{6}$. Hence we can add it to $\frac{5}{6}$, making $\frac{9}{6}$, which is then simplified to $1\frac{1}{2}$. Another example is

$$\frac{1}{3} + \frac{5}{6} + \frac{7}{8} = \frac{8}{24} + \frac{20}{24} + \frac{21}{24} = \frac{8 + 20 + 21}{24}$$

$$= \frac{49}{24} = 2\frac{1}{24}$$

It is generally preferable to divide out fractions greater than 1 as we have done above.

(d) *Subtract*
Exactly the same rules apply to the subtraction of fractions.

We can also express parts of the whole as "decimals" or $\frac{1}{10}$ parts, written as 0·1, 0·2, 0·3 etc (these are equivalent to $\frac{1}{10}$, $\frac{2}{10}$, $\frac{3}{10}$ etc). The "full stop" is known as the *decimal point*. In a decimal, the "nought" before the decimal point should never be omitted.

The denominator of any fraction can be divided into the numerator to give a decimal, eg

$$\frac{1}{8} = 0 \cdot 125$$
$$\frac{3}{8} = 0 \cdot 375$$

The more common fractions and decimal equivalents should be memorized, eg

$$\frac{1}{10} = 0 \cdot 1 \qquad \frac{1}{8} = 0 \cdot 125$$

$$\frac{2}{10} = \frac{1}{5} = 0 \cdot 2 \qquad \frac{2}{8} = \frac{1}{4} = 0 \cdot 25$$

$$\frac{3}{10} = 0.3 \qquad \frac{3}{8} = 0.375$$

$$\frac{4}{10} = \frac{2}{5} = 0.4 \text{ etc} \qquad \frac{4}{8} = \frac{1}{2} = 0.5 \text{ etc}$$

Numbers can be expressed to "so many significant figures" or "so many decimal places".

Thus 12345 is a number to five significant figures
1234 is a number to four significant figures
123 is a number to three significant figures

Note also 1·23 is a number to three significant figures (the decimal point is ignored).

12·345 is a number to three decimal places
12·34 is a number to two decimal places
12·3 is a number to one decimal place

Decimals may be "rounded off", that means

3·3267 to three decimal places is 3·327
(the 7 is greater than 5, so 6 becomes 7)
3·327 to two decimal places is 3·33
(the 7 is greater than 5, so 2 becomes 3)
3·33 to one decimal place is 3·3
(the 3 is less than 5, so is ignored)

Powers of numbers

When, say, two of a certain number are multiplied together, that number is said to be "raised to the power 2". Thus $2 \times 2 = 4$ means that 2 raised to the power 2 is 4. In this case we would say 2 "squared" is 4 and write it as $2^2 = 4$. The "little 2 up in the air" is called an *index*. Similarly $2 \times 2 \times 2 = 8$ means that 2 raised to the power 3 is 8, or 2 "cubed" is 8, written as $2^3 = 8$. Also $2 \times 2 \times 2 \times 2 = 16$. Here we have no alternative but to say 2 "to the power 4" $= 16$ or $2^4 = 16$.

The use of indices or the index notation is a very convenient way of expressing the large numbers which often occur in radio calculations, eg

$$100 = 10 \times 10 = 10^2$$
$$10,000 = 10 \times 10 \times 10 \times 10 = 10^4$$
$$1,000,000 = 10 \times 10 \times 10 \times 10 \times 10 \times 10 = 10^6$$

Note that $10 = 10^1$ (the index here is taken for granted). Similarly

$$\frac{1}{100} = \frac{1}{10 \times 10} = \frac{1}{10^2} \text{ (written as } 10^{-2})$$

$$\frac{1}{10,000} = \frac{1}{10 \times 10 \times 10 \times 10} = \frac{1}{10^4} \text{ (written as } 10^{-4})$$

$$\frac{1}{1,000,000} = \frac{1}{10 \times 10 \times 10 \times 10 \times 10 \times 10} = \frac{1}{10^6}$$
(written as 10^{-6})

Numbers expressed in the index notation are multiplied and divided by adding and subtracting respectively the indices.

$$10^2 \times 10^3 = 10^{2+3} = 10^5$$

$$10^4 \div 10^2 = 10^{4-2} = 10^2$$
$$\frac{10^5 \times 10^7 \times 10^{-2}}{10^3 \times 10 \times 10^{-3}} = \frac{10^{5+7-2}}{10^{3+1-3}} = \frac{10^{10}}{10^1} = 10^9.$$

We can do this as long as the "base" is the same in each case. In the above examples, the "base" is 10. For example $10^2 \times 2^2 = 100 \times 4 = 400$, which is neither 10^4 or 2^4!

Roots of numbers

The root of a number is that number which when multiplied by itself so many times equals the given number; the "square" root of 4 is 2, ie $2 \times 2 = 4$, and this is written $^2\sqrt{4} = 2$.

Similarly the "cube" root of 8 is 2, ie $2 \times 2 \times 2 = 8$, and $^4\sqrt{16} = 2$ etc. Note the little 2 in the sign for square root is normally omitted so that $\sqrt{}$ signifies the square root.

Numbers like 4, 16 and 25 are called *perfect* squares because their square roots are whole numbers, thus

$$\sqrt{49} = 7 \qquad \sqrt{121} = 11 \quad \text{etc}$$

The following should be memorized as they can often be very useful.

$$\sqrt{2} = 1.41 \qquad \sqrt{3} = 1.73 \qquad \sqrt{5} = 2.24 \qquad \sqrt{10} = 3.162$$

For example

$$\sqrt{200} = \sqrt{2 \times 100} = \sqrt{2} \times \sqrt{100} = 1.41 \times 10 = 14.1$$
$$\sqrt{192} = \sqrt{3 \times 64} = \sqrt{3} \times \sqrt{64} = 1.73 \times 8 = 13.8$$

It is always worth checking to see if the number left after dividing by 2, 3 or 5 is a perfect square.

The square root of a number expressed in the index notation is found by dividing the index by 2, thus $\sqrt{10^6} = 10^3$ and $\sqrt{10^{12}} = 10^6$ and so on. Similarly $\sqrt{10^{-6}} = 10^{-3}$ etc. Should the index be an odd number, it must be made into an even number as follows.

$$\sqrt{10^{-15}} = \sqrt{10 \times 10^{-16}}$$

$$= \sqrt{10} \times 10^{-8}$$
$$= 3.162 \times 10^{-8}$$

The constant term "π" occurs in many calculations; "π" has great significance in mathematics and is, the ratio of the circumference and the diameter of a circle. π can be taken to be 3·14 or 22/7. The error in taking π^2 as 10 is less than 1·5 per cent and is acceptable here. $1/\pi$ can be taken as 0·32 and $1/2\pi$ as 0·16 (the error in calling this $\frac{1}{6}$ is really somewhat too high). $1/2\pi = 0.16$ is particularly useful.

Typical calculations

We will now apply these rules to the solution of problems likely to be met in radio work as a lead-in to some typical numerical multiple-choice questions.

Answers to three significant figures as given by a slide

rule or four-figure logarithm tables are satisfactory for most radio purposes and the eight figures given by the electronic calculator should certainly be rounded off.

The most important aspect is to remember that the units met with are most likely to be the practical ones such as microfarads, picofarads, milliamperes, millihenrys etc. These must be converted to the basic units of farads, amperes and henrys before substituting them into the appropriate formula. This involves multiplying or dividing by 1,000 (10^3), 1,000,000 (10^6) and so on. Therefore the important thing is to get the decimal point in the right place or the right number of noughts in the answer. The commonest conversions are as follows:

There are 10^6 microfarads in 1 farad,
    hence $8\mu F = 8 \times 10^{-6}$ farads
There are 10^{12} picofarads in 1 farad,
    hence $22pF = 22 \times 10^{-12}$ farads.

(The use of "nano" or 10^{-9} is now fairly common; there are 10^9 nanofarads in 1 farad so $1nF = 1 \times 10^{-9}$ farads, but such a capacitor may well be marked "1,000pF".) Similarly other conversions are

$$50\mu H = 50 \times 10^{-6} \text{ henrys}$$
$$3mH = 3 \times 10^{-3} \text{ henrys}$$
$$45mA = 45 \times 10^{-3} \text{ amperes}$$
$$10\mu A = 10 \times 10^{-6} \text{ amperes}$$

Problem 1
What value of resistor is required to drop 150V when the current flowing through it is 25mA?

This involves Ohm's Law which can be expressed in symbols in three ways:

$$R = \frac{V}{I} \qquad I = \frac{V}{R} \qquad V = I \times R$$

where R is in ohms, V in volts and I in amperes. Clearly the first, $R = V/I$, is needed. First of all, we must express the current (25mA) in amperes.

$$25mA = \frac{25}{1,000}A \text{ (or } 25 \times 10^{-3}A)$$

Substituting values for V and I

$$R = \frac{V}{I}$$
$$= 150 \times \frac{1,000}{25}$$

(we are dividing by $\frac{25}{1,000}$, ie multiplying by $\frac{1,000}{25}$)

hence
$$R = \frac{150 \times 1,000}{25}$$
25 "goes into" 150 six times, so
$$R = 6 \times 1,000$$
$$= 6,000\Omega$$

Problem 2
What power is being dissipated by the resistor in Problem 1?

The power dissipated in the resistor is power (watts) = V (volts) $\times I$ (amps). By Ohm's Law, power can be expressed in two other forms.

$$W = \frac{V^2}{R} \text{ and } W = I^2R$$

because we know V, I and R we can use any of the above relationships, say

$$W = \frac{V^2}{R}$$
$$W = \frac{150 \times 150}{6,000}$$

Two "noughts" on the top and the bottom can be cancelled, leaving

$$= \frac{15 \times 15}{60}$$

Cancelling 15 into 60 leaves

$$= \frac{15}{4} = 3\tfrac{3}{4} \text{ W}$$

The other two forms will, of course, give the same answer—try them!

Problem 3
Resistors of 12Ω, 15Ω and 20Ω are in parallel. What is the effective resistance?

$$\frac{1}{R} = \frac{1}{R_1} + \frac{1}{R_2} + \frac{1}{R_3}$$
$$= \frac{1}{12} + \frac{1}{15} + \frac{1}{20}$$

60 is the lowest common denominator of 12, 15 and 20, so

$$\frac{1}{R} = \frac{5}{60} + \frac{4}{60} + \frac{3}{60}$$
$$= \frac{5 + 4 + 3}{60}$$
$$= \frac{12}{60}$$

This is a simple equation in R, and the first step in solving it is to "cross-multiply". It may be shown that the denominator of one side multiplied by the numerator of the other side is equal to the numerator of the first side multiplied by the denominator of the other side, thus

$$R \times 12 = 1 \times 60$$

Hence, dividing each side by 12

$$R = \frac{60}{12}$$
$$R = 5\Omega.$$

Problem 4
Capacitors of 330pF, 680pF and 0·001μF are in parallel. What is the effective capacitance?

The first step is to express all the capacitors in the *same* units which can be either picofarads or microfarads.

$$0·001μF = 0·001 × 1,000,000pF$$

(there are 1,000,000pF in 1μF) and hence

$$0·001μF = 1,000pF.$$

Effective capacitance is therefore

$$330pF + 680pF + 1,000pF = 2,010pF.$$

Problem 5
What is the reactance of a 30H smoothing choke at a frequency of 100Hz?

$$X_L = 2πfL$$
$$X_L = 2π × 100 × 30Ω$$
$$= 6,000π \text{ ohms}$$

We take $π$ to be 3·14 so

$$X_L = 6,000 × 3·14$$
$$= 18,840Ω$$

Problem 6
What is the reactance of a 100pF capacitor at a frequency of 20MHz?

$$X_C = \frac{1}{2πfC}$$

(X_C is in ohms when f is in hertz and L in henrys)

$$f = 20MHz = 20 × 10^6Hz = 2 × 10^7Hz$$
$$C = 100pF = 100 × 10^{-12}F = 10^{-10}F$$

(it is much more convenient here to use the index notation) hence

$$X_C = \frac{1}{2π × 2 × 10^7 × 10^{-10}} \text{ ohms}$$

$$= \frac{1}{2π} × \frac{1}{2 × 10^{-3}}$$

Note that we have kept $\frac{1}{2π}$ intact because $\frac{1}{2π} = 0·16$, thus

$$X_C = 0·16 × \frac{1}{2 × 10^{-3}}$$

$$= \frac{0·16 × 1,000}{2}$$

$$= 80Ω$$

Problem 7
What is the impedance (Z) of an inductance which has a resistance (R) of 4Ω and a reactance (X) of 3Ω?

$$Z = \sqrt{(R^2 + X^2)}$$
$$= \sqrt{(4^2 + 3^2)}$$
$$= \sqrt{16 + 9}$$
$$= \sqrt{25}$$
$$= 5Ω$$

Problem 8
At what frequency do a capacitor of 100pF and an inductance of 100μH resonate?

At resonance

$$2πfL = \frac{1}{2πfC}$$

hence

$$f = \frac{1}{2π\sqrt{LC}}$$

(f is in hertz, L is in henrys, C is in farads)

$$100μH = 100 × 10^{-6}H$$
$$100pF = 100 × 10^{-12}F$$

$$f = \frac{1}{2π\sqrt{LC}}$$

$$= \frac{1}{2π\sqrt{100 × 10^{-6} × 100 × 10^{-12}}}$$

$$= \frac{1}{2π\sqrt{10^2 × 10^{-6} × 10^2 × 10^{-12}}}$$

$$= \frac{1}{2π\sqrt{10^{-14}}}$$

$$= \frac{1}{2π × 10^{-7}}$$

$$= \frac{1}{2π} × 10^7$$

$$= 0·16 × 10^7$$

$$= 1·6 × 10^6Hz$$

$$= 1·6MHz$$

Numerical multiple-choice questions in the RAE

The numerical multiple-choice questions set in the RAE involve quite simple calculations in order to decide which of the four answers given is correct. The questions are likely to be similar to the problems just worked through and generally the answer comes out without the need for any aid to calculation. As in solving the previous problems, the most important thing is to "get the units right". The way to solve these questions should be clear from the following worked examples.

Question 1

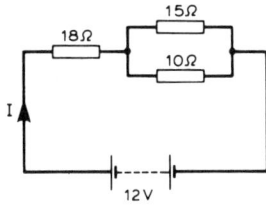

The current I is
  (a) 0·25A.
  (b) 0·43A.
  (c) 0·5A.
  (d) 0·67A.

The effective resistance of the two resistors in parallel is

$$R_{eff} = \frac{15 \times 10}{25} = 6\Omega$$

The effective resistance of the whole circuit is

$$R_{eff} = 18 + 6 = 24\Omega$$

$$I = \frac{12}{24} = 0·5A$$

Answer (c) is therefore correct.

Question 2

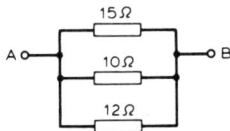

The effective resistance between points A and B is
  (a) 4Ω.
  (b) 6Ω.
  (c) 17Ω.
  (d) 37Ω.

The effective resistance must have a value less than the value of the smallest resistor, so neither answers (c) nor (d) are correct. Take the top two resistors and apply the formula

$$R_{eff} = \frac{R_1 \times R_2}{R_1 + R_2} = \frac{15 \times 10}{25} = 6\Omega$$

Again apply the formula to include the 12Ω resistor.

$$R_{eff} = \frac{6 \times 12}{18} = \frac{72}{18} = 4\Omega$$

Answer (a) is therefore correct.

Question 3

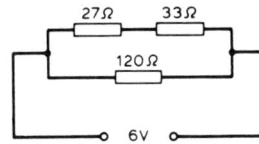

The current flowing through the 27Ω resistor has a value of
  (a) 27mA.
  (b) 33mA.
  (c) 60mA.
  (d) 100mA.

The current flowing through the 120Ω resistor has no bearing on the answer. The current through the 27Ω resistor will be the same as that through the 33Ω resistor. The current through the two resistors in series

$$= \frac{6}{27 + 33} = \frac{6}{60} = \frac{1}{10} \text{ A}$$

The correct answer is (d).

Question 4
The voltage across the 33Ω resistor in the previous question is
  (a) 0·6V.
  (b) 1·2V.
  (c) 3·3V.
  (d) 4·5V.

$$V = I \times R = \frac{1}{10} \times 33 = 3·3V$$

The correct answer is (c).

Question 5
A λ/2 dipole has a length of just under 7·5m. It will be resonant at a frequency of approximately
  (a) 15MHz.
  (b) 20MHz.
  (c) 25MHz.
  (d) 30MHz.

$$\lambda = 15m \quad f = \frac{c}{\lambda} = \frac{300 \times 10^6}{15} = 20 \times 10^6 Hz = 20MHz.$$

Therefore (b) is the correct answer.

Question 6
An oscilloscope shows the peak-to-peak voltage of a sine wave to be 100V. The rms value is
  (a) 27·28V.
  (b) 35·35V.
  (c) 50V.
  (d) 70·7V.

$$V_{rms} = V_{peak} \times 0·707$$
$$= 50 \times 0·707 = 35·35V$$

The correct answer is (b).

Question 7

The internal capacitance between the base and emitter of a transistor is 2pF. The reactance at a frequency of 500MHz will be approximately

(a) 16Ω.
(b) 160Ω.
(c) 1·6kΩ.
(d) 16kΩ.

$$X_C = \frac{1}{\omega C} = \frac{1}{2\pi \times 500 \times 10^6 \times 2 \times 10^{-12}} = \frac{1}{2\pi \times 10^{-3}}$$

$$= 0·16 \times 10^3 = 160\Omega$$

The correct answer is (b).

Question 8

A loudspeaker speech coil has a resistance of 3Ω. If the voltage across it is 3V, then the power in the speech coil is

(a) 1W.
(b) 3W.
(c) 6W.
(d) 9W.

$$P = \frac{V^2}{R} \text{ watts} \qquad P = \frac{3^2}{3} = 3W$$

Answer (b) is therefore correct.

Question 9

A smoothing choke has an inductance of 0·2H. Its reactance at a frequency of 100Hz is approximately

(a) 40Ω.
(b) 125Ω.
(c) 400Ω.
(d) 1250Ω.

$$X_L = 2\pi fL = 2\pi \times 100 \times 0·2 = 40\pi \quad \text{or about } 125\Omega.$$

Hence (b) is the correct answer.

Question 10

A coil has a reactance of 1,000Ω and a resistance of 10Ω. Its approximate impedance is

(a) 990Ω.
(b) 1000Ω.
(c) 1100Ω.
(d) 10kΩ.

$$Z = \sqrt{R^2 + X_L^2} = \sqrt{100 + 10^6} = \sqrt{1,000,100} \approx 1,000\Omega$$

The effect of the resistance is so small that it can be neglected, so (b) is the correct answer.

Question 11

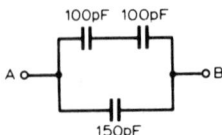

The capacitance measured between terminals A and B will be

(a) 37·5pF.
(b) 50pF.
(c) 200pF.
(d) 350pF.

The capacitance must be greater than 150pF. The two 100pF capacitors in series have an effective capacitance of 50pF. Therefore the answer is 150 + 50 = 200pF, ie answer (c).

Question 12

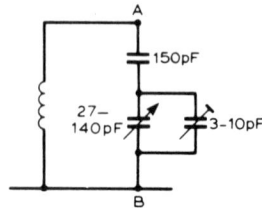

When the variable capacitor and the trimmer capacitor of a local oscillator tuned circuit are adjusted to their maximum values, the effective capacitance between points A and B will be

(a) 50pF.
(b) 75pF.
(c) 200pF.
(d) 300pF.

The tuning and trimmer capacitors will have an effective capacitance of 140 + 10 = 150pF. Therefore capacitance between A and B will be 75pF, and (b) is the correct answer.

Question 13

When the variable capacitor and the trimmer are set at minimum, the effective capacitance between points A and B will be

(a) 25pF.
(b) 75pF.
(c) 120pF.
(d) 180pF.

Using the same calculations, (a) is the correct answer.

Question 14

The peak-to-peak value of a sine wave having an rms voltage of 14·1V is approximately

(a) 20V.
(b) 28·2V.
(c) 40V.
(d) 56·4V.

The peak value for the positive half-cycle is 14·1 × √2 = 20V. Therefore the peak-to-peak voltage = 2 × 20 = 40V.

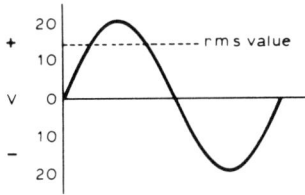

(c) is the correct answer.

Question 15

A quarter-wave antenna is resonant at 10MHz. Its approximate length will be

(a) 7·5m.
(b) 15m.
(c) 20m.
(d) 30m.

$$\lambda = \frac{C}{f} = \frac{300 \times 10^6}{10 \times 10^6} = 30m$$

$$\lambda/4 = 7·5m$$

(a) is the correct answer.

Summary of formulae

Ohm's Law $R = \dfrac{V}{I}$ $V = IR$ $I = \dfrac{V}{R}$

Power $W = V \times I$ $W = I^2R$ $W = \dfrac{V^2}{R}$

Reactance $X_L = 2\pi fL$

$$X_C = \frac{1}{2\pi fC}$$

Resonance $f = \dfrac{1}{2\pi \sqrt{LC}}$

Resistors (series) $R = R_1 + R_2 + R_3 + \ldots$

Resistors (parallel) $\dfrac{1}{R} = \dfrac{1}{R_1} + \dfrac{1}{R_2} + \dfrac{1}{R_3} + \ldots$

Capacitors (series) $\dfrac{1}{C} = \dfrac{1}{C_1} + \dfrac{1}{C_2} + \dfrac{1}{C_3} + \ldots$

Capacitors (parallel) $C = C_1 + C_2 + C_3 + \ldots$

Wavelength (metres) $= \dfrac{300}{f(\text{MHz})}$

For a sine wave, rms value = $0·707 \times$ peak value

CHAPTER 4

Preparing for the RAE

Most RAE candidates take a part-time evening course run by their local authority. Lists of colleges and other institutions, which offer a course leading to the RAE, together with enrolment and starting dates, are given in *Radio Communication* and other amateur radio magazines, generally in the July, August and September issues. However, if the local college is not mentioned, it does not necessarily mean that an RAE course is not available there and enquiries should be made directly.

Normally, college regulations require a minimum of 12 students before a course can be run. It is therefore advisable to enrol for the course as early as possible so that the college is aware that there are likely to be sufficient students to justify the course being held.

Courses usually start in late August or early September and aim at the examination which is held in the following May. The examination is also held in December.

It should be noted that most technical and other colleges do not usually accept students below the age of 16. But students who are still at school are accepted provided that they are over 16 and in their last year at school. The agreement of their headmaster and parents is necessary. In many colleges, acceptance is also at the discretion of the lecturer.

If there is no course available at the local college, do not give up hope! Enquiries, via the local newspaper, for instance, may well reveal a sufficient number of likely students to interest the college in the possibility of starting a course.

Enquiry of the Membership Services Department of the RSGB should lead to the location of the nearest course, but there are inevitably areas of the UK where the nearest RAE course may be beyond convenient travelling distance.

Some amateur radio clubs organise an RAE course, so contact with the secretary of the local club may well be worthwhile.

In this situation, do not forget that many enthusiasts study on their own for the RAE using the *RAE Manual* published by the RSGB. It is also essential to obtain the latest edition the booklet *How to become a radio amateur* in order to be familiar with the current licence conditions. This is free and may be obtained from: The Amateur Radio Licensing Unit, Post Office Headquarters, Chetwynd House, Chesterfield S49 1PF.

There are a number of correspondence courses available but the RSGB is not able to recommend any particular one.

Details of the examination

The regulations for the RAE may be obtained from: City and Guilds of London Institute, 46 Britannia Street, London WC1X 9RG. In this pamphlet, details of the examination itself are presented as examination objectives. These describe in general terms the nature of the questions and the syllabus outlines the subject matter to which the questions relate.

Entry for the examination must be made on the appropriate form. This has to be submitted to the examination centre, usually the local college, by the date required and accompanied by the appropriate fee.

Some colleges will accept external candidates for the RAE even though they do not hold an RAE course.

If you are having difficulty in finding a centre, the Membership Services Department of the RSGB will be able to help.

Background knowledge of amateur radio

It is most important that the RAE candidate should acquire as much background knowledge of amateur radio as possible. A period as a shortwave listener is particularly valuable. One then becomes familiar with amateur radio communication, how propagation governs which part of the world can be heard, when and on which waveband, operating procedures and so on. A number of journals wholly concerned with amateur radio are now available, these include *Radio Communication* published by the RSGB. While every article in these is obviously not aimed at the beginner, much useful information can be found in them. If there is a local radio society, join it! There one meets other amateurs, some with wide experience and some just beginners. Talking to them and listening to their conversations can be most helpful.

Basic examination requirements

The RAE is held in two parts:

Paper 1 (765-1-01): Licensing conditions, transmitter interference and electromagnetic compatibility.

Paper 2 (765-1-02): Operating procedures, practices and theory

There is a break of 15 minutes between the two papers.

Candidates must take both papers on their first entry, but those who are successful in only one part may carry forward their success and hence need only re-take the part in which they were unsuccessful.

Because the questions in the RAE are multiple-choice, it does not mean that knowledge of circuit diagrams of receiver and transmitter stages is not required. Study those in the *RAE Manual* and compare the values of components in various circuits. Be familiar with the modern circuit symbols for the common electronic components. These are in accordance with British Standard 3939 (and are given in appendix 1 of the *RAE Manual*).

A number of multiple-choice questions are numerical. While these can be ignored with a fair degree of safety by the candidate who is not strong in mathematics, the standard of mathematics involved is quite low and they should be attempted if at all possible (chapter 3 gives information on mathematics for the RAE.)

These questions generally involve the calculation of the total resistance or capacitance of several resistors or capacitors in series or parallel; the calculation using Ohm's Law of current or voltage in a simple circuit or the manipulation of the formulas for inductive or capacitive reactance and resonance frequency. These normally involve the insertion in the formulas of factors such as 10^{-12} or 10^{-6} to convert practical units like $\mu\mu F(pF)$ to the basic unit (F) or μH to H.

In the examination room

It is most important to find out exactly where the examination is to be held and to arrive there 10 minutes or so before the examination commences.

There are a number of formalities to go through. The examination regulations have to be read to the candidates by the invigilator. Your lecturer may well be present but he is unlikely to be the invigilator. Your centre number and candidate number will have been notified to you by the college authorities. These numbers together with your name and address and other information have to be written down in several places. This is obviously vitally important!

In the examination you are required to indicate which of the four possible answers you consider to be the correct one by filling the appropriate box a, b, c or d on the answer sheet with an HB pencil. The latter is most important: an HB pencil is provided. The following illustrations show the front cover of the question book and the answer sheet. From these you will see what information you have to insert. In particular you must indicate your candidate number by filling in the numbered boxes at the top right-hand corner of the sheet. Thus your answers and identification number can be scanned by the optical character reader.

Any rough work or calculations can be done on the question book. Note that the question book cannot be taken away: it is a C & G requirement that it be handed in at the end of the examination and returned to them.

Do remember the following

Candidates are not competing against each other and are only endeavouring to reach a certain standard.

Don't imagine that the examination is so easy that one cannot fail or on the other hand, feel that the questions are bound to be beyond one's capability. You cannot fool the computer by filling all the spaces provided for each answer. Neither can you pass by filling in answers at random.

A great many candidates of all ages from all walks of life pass this examination. Finally good luck and don't forget to collect your certificate from the college!

MA

City and Guilds of London Institute 46 BRITANNIA STREET, LONDON, WC1X 9RG

Candidate Name		Date	Paper Number

Centre Name		Centre Number	Candidate Number

Fill in your
answers in
the boxes
—

WORK

DOWN

THE

COLUMNS

1	26	51	76
2	27	52	77
3	28	53	78
4	29	54	79
5	30	55	80
6	31	56	81
7	32	57	82
8	33	58	83
9	34	59	84
10	35	60	85
11	36	61	86
12	37	62	87
13	38	63	88
14	39	64	89
15	40	65	90
16	41	66	91
17	42	67	92
18	43	68	93
19	44	69	94
20	45	70	95
21	46	71	96
22	47	72	97
23	48	73	98
24	49	74	99
25	50	75	100

Now go back and check your work

Illustration of RAE answer sheet

CHAPTER 5

Practice multiple choice questions

This chapter contains nine sample multiple choice examination papers in the pattern of the RAE.

These papers are closely representative of the scope of the examination in content and difficulty but not exactly so as they were not originated by the City and Guilds of London Institute.

Paper 1 should be completed in 1 hour and 15 minutes. After a break of 15 minutes, Paper 2 should be completed in 1 hour and 30 minutes. Answers will be found at the end of the chapter.

Sample examination 1, Paper 1
Licensing conditions, transmitter interference and electromagnetic compatibility.

1. If the licence is revoked it shall be:
   a) just destroyed
   b) returned to the Licensing Authority
   c) returned to any Post Office
   d) returned to the RSGB

2. The class B licence does not authorise the use of frequencies:
   a) above 144MHz
   b) above 430MHz
   c) in the microwave range
   d) below 30MHz

3. Providing the licence fee is paid annually the licence is valid:
   a) 5 years
   b) 10 years
   c) 25 years
   d) until revoked

4. The callsign GI0xxx is issued to an amateur living in:
   a) England
   b) Wales
   c) Northern Ireland
   d) Jersey

5. Which of the following is permissible for a log?
   a) a computer print out on perforated sheets
   b) a loose leaf notebook
   c) a writing pad
   d) a stapled exercise book

6. Slow scan and high definition television using frequency modulation is:
   a) A3F
   b) C3F
   c) F3E
   d) F3F

7. The minimum age when the RAE can be taken is:
   a) not specified
   b) 12 years
   c) 13 years
   d) 14 years

8. If a station is located 0.3km from the boundary of an airfield, the height of the antenna system must not exceed:
   a) 10 metres
   b) 15 metres
   c) 20 metres
   d) 50 metres

9. When using equipment travelling in a car the suffix to be used is:
   a) /P
   b) /C
   c) /M
   d) not defined

10. Times entered in the log must be in:
    a) UTC
    b) local time
    c) the local time of the other station
    d) BST

11. CQ calls must:
    a) always be in cw
    b) never be entered in the log
    c) always be entered in the log
    d) never be sent

12. The nationality requirement for holding an amateur licence in the UK is:
    a) British
    b) British and Commonwealth
    c) European
    d) not specified

13. According to the UK Amateur Licence, third party traffic is:
    a) prohibited in the UK
    b) permitted with the USA
    c) allowed in the UK under certain restricted circumstances
    d) permitted at all times

14. The 24.89 - 24.99MHz band is restricted to:
    a) vertical polarisation only
    b) A1A keying only
    c) F3E only
    d) rtty

15. Terms of the Amateur Licence can be varied by a

general notice in which of the following?
a) London, Glasgow and Cardiff Gazettes
b) Glasgow, Cardiff and Belfast Gazettes
c) Edinburgh, Belfast and London Gazettes
d) Cardiff, Edinburgh and Manchester Gazettes

16. Tests to ensure that no undue interference is being caused should be carried out:
a) from time to time
b) every three months
c) every six months
d) every year

17. Which of the following filters would minimise harmonic output from an hf transmitter?

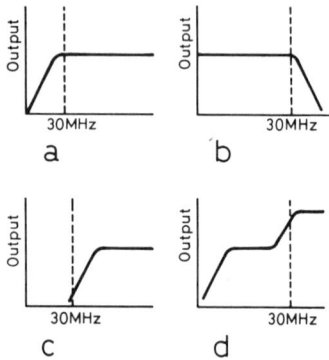

18. An absolutely pure sine wave emission:
a) can never cause interference
b) contains high harmonic content
c) contains a high level of spurious signals
d) could cause interference problems

19. The first odd harmonic of 144.69MHz is:
a) 48.23MHz
b) 289.38MHz
c) 434.07MHz
d) 723.45MHz

20. An 8MHz crystal oscillator used in a transmitter is followed by several multiplier stages, the first few of which are ×2,×2,×3. If not carefully screened these could cause interference on a nearby:
a) 3.5MHz receiver
b) 88-108MHz fm broadcast receiver
c) 10MHz transmitter
d) none of these

21. Any non-linear device will produce:
a) mixing products
b) amplification
c) filtering
d) key clicks

22. Which of the following is likely to give minimum interference due to key clicks?

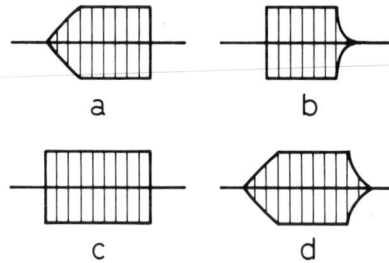

23. To maintain good screening of equipment with hinged lids it is advisable to connect across the hinge:
a) a plastic retainer
b) a good earth strap
c) a polystyrene capacitor
d) a ferrite bead

24. Which of the following operating conditions of the power amplifier stage of a transmitter is likely to produce the highest harmonic content in the output waveform?
a) Class C
b) Class B
c) Class AB
d) Class A

25. An absorption wavemeter can be used to check for:
a) over-modulation
b) receiver overloading
c) band edge signals
d) correct selection of harmonic from a multiplier circuit

26. A broadband transistor p.a should be followed by:
a) a high pass filter
b) a low pass filter
c) a resistive attenuator
d) a mains filter

27. In order to minimise splatter, the audio bandwidth should be restricted to:
a) 1kHz
b) 1.5kHz
c) 2kHz
d) 3kHz

28. A bandpass filter following a vhf transmitter will:
a) stop all transmitting frequencies
b) allow all harmonics to be radiated
c) allow all sub-harmonics to be radiated
d) pass the desired frequency range with minimum loss

29. Spurious oscillations may be caused by:
a) self resonance of a carbon resistor
b) self resonance in a diode
c) self resonance of an rf choke
d) damping

30. Which of the circuits below would provide low pass filtering for a microphone circuit in order to minimise bandwidth?

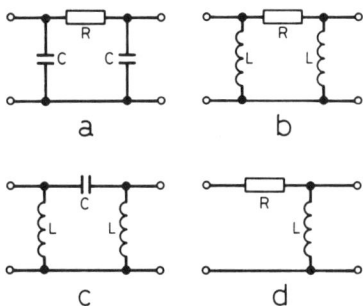

31. A neighbour's hi-fi system is suffering rf breakthrough. One possible cure would be:
a) ferrite bead on the transmitter lead
b) a capacitor across the transmitter lead
c) screened wire for the loudspeaker leads
d) open wire feeder for the transmitter lead

32. Which of the following is most likely to produce broad band continuous interference?
a) an electric light switch
b) an incandescent bulb
c) a microwave transmitter
d) poor commutation in an electric drill

33. The sensitivity of a receiver can be degraded by:
a) strong rf signals on a nearby frequency
b) removing all crystals
c) good af filtering
d) incorrect adjustment of the volume control

34. Before explaining to someone that the cause of interference is lack of immunity in their equipment, one should:
a) make sure that there is no breakthrough on one's own domestic equipment
b) disconnect all one's equipment from the mains
c) write a letter to the DTI
d) ignore all complaints

35. An amateur radio transmitter/antenna system has an erp of 100 watts, the field strength at a distance of 100 metres is:
a) 0.35V/m
b) 0.7V/m
c) 3.5V/m
d) 7.02V/m

36. Rectification of an rf signal in an audio amplifier is likely to occur:
a) in a tantalum capacitor
b) at a base-emitter junction
c) at the junction of two resistors
d) in a copper wire

37. All equipment carrying rf currents should be:
a) not earthed
b) connected to the mains
c) screened as well as possible
d) be left "floating"

38. When living in a densely populated area and during evening tv hours it is advisable to:
a) always use maximum transmitter output
b) use only sufficient power to maintain communications
c) use bands that are known to cause tvi
d) ignore all complaints

39. A braid-breaking choke in a tv antenna downlead will block:
a) all ac signals
b) out of phase interfering signals
c) in phase interfering signals
d) mains hum

40. Capacitors to be used in rf filters should be:
a) aluminium electrolytics
b) tantalum electrolytics
c) ceramic
d) polycarbonate

41. The typical wideband tv preamplifier is susceptible to:
a) mains hum
b) dc supply variations
c) overloading from a remote transmitter
d) overloading from a nearby transmitter

42. The third harmonic from a 29MHz transmission lies in:
a) a uhf band
b) the fm broadcast band
c) a pmr band
d) another amateur band

43. Unwanted rf pick up in the i.f stage of a tv set usually results in:
a) problems with the tv picture
b) poor power supply regulation
c) random channel changing
d) no audio output

44. To reduce strong signals from a 21MHz transmitter reaching a tv via the antenna downlead, one could fit:
a) a high pass filter in the tv downlead
b) a low pass filter in the tv downlead
c) a uhf amplifier in the tv downlead
d) a band reject filter at the tv channel frequency

45. If an antenna runs close and parallel to an overhead 240V ac power line, there may be the possibility of:

a) cheap power
b) harmonic generation
c) producing mains borne interference
d) 50Hz modulation on all signals

Sample examination 1, Paper 2
Operating practices, procedures and theory.

1. To prevent annoying other users on a band, a transmitter should always be tuned initially:
    a) on a harmonic
    b) into an antenna
    c) into a dummy load
    d) on a dipole

2. COIL using the International Phonetic Alphabet would be:
    a) charlie, ocean, italy, lima
    b) charlie, oscar, india, lima
    c) coil, oscar, inductance, london
    d) charlie, oscar, india, london

3. The band plans should be observed because:
    a) they are mandatory
    b) they are governed by international regulations
    c) they aid operating
    d) they are only for novices

4. The Q code for changing frequency is:
    a) QSF
    b) QSY
    c) QRF
    d) QCF

5. The difference in frequency between input and output of a 432MHz repeater in the UK is:
    a) 600kHz
    b) 1.6MHz
    c) 2.4MHz
    d) 4.5MHz

6. In a cw call the abbreviation KN means:
    a) any station to reply
    b) this is the end of a test transmission
    c) the reply is expected in telephony
    d) reply only expected from the station called

7. A readability report R4 indicates:
    a) unreadable
    b) readable with considerable difficulty
    c) readable with practically no difficulty
    d) perfectly readable

8. When using a repeater, priority should be given to:
    a) stations operating /M
    b) dx stations
    c) members of the local repeater group
    d) base stations

9. In a cw contact WX refers to:
    a) working conditions
    b) weather
    c) wife
    d) type of antenna

10. The effective resistance of three 24 ohms resistors connected in parallel is:
    a) 8 ohms
    b) 12 ohms
    c) 36 ohms
    d) 72 ohms

11. A sinewave has an rms value of 12V, the peak to peak value of this is:
    a) 16.97V
    b) 24V
    c) 33.9V
    d) 36.4V

12. What is the Q factor of a filter if the centre frequency is 9MHz and the half bandwidth is 1.5kHz ?
    a) 6
    b) 300
    c) 3000
    d) 6000

13. A power gain of 4 is equivalent to:
    a) 3dB
    b) 6dB
    c) 10dB
    d) 16dB

14. A period of $50\mu s$ corresponds to:
    a) 2kHz
    b) 20kHz
    c) 200kHz
    d) 2MHz

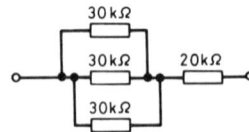

15. The total resistance of the above circuit is:
    a) 5 kohm
    b) 15 kohm
    c) 20 kohm
    d) 30 kohm

16. If 10mA is to flow through the above combination, the voltage of A with respect to B must be:
    a) −10.6V
    b) −10V

c) 10V
d) 10.6V

17. In a forward biased pn junction, the electrons:
a) flow from p to n
b) flow from n to p
c) remain in the n region
d) disintegrate

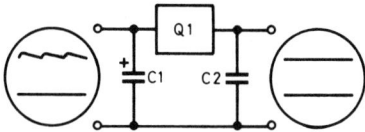

18. The above circuit is part of a power supply. The oscilloscope traces give the waveform at two points in the circuit. Device Q1 is:
a) a bridge rectifier
b) a voltage stabiliser
c) a single diode
d) a varactor diode

19. The purpose of C1 in the above question is:
a) current limiting
b) rf coupling
c) ac coupling
d) smoothing

20. The voltage drop across a germanium diode when conducting is about:
a) 0.3V
b) 0.6V
c) 0.7V
d) 1.3V

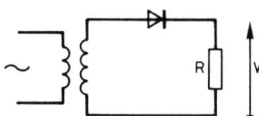

21. The above circuit is for:
a) voltage stabilisation
b) rectification
c) reverse bias protection
d) voltage multiplication

22. In the circuit of Q21, which of the following represents the output waveform?

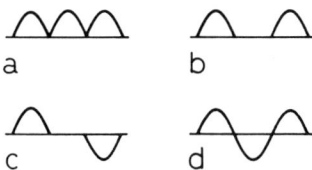

23. AGC stands for:
a) amplified gain control
b) auxiliary gain cut-off
c) automatic ganging control
d) automatic gain control

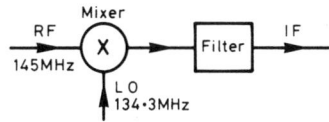

24. In the above arrangement, second channel interference could be experienced at:
a) 123.6MHz
b) 134.3MHz
c) 145MHz
d) 155.7MHz

25. The bandwidth of the filter in Q24 for the reception of a.m signals would be:
a) 3kHz
b) 6kHz
c) 10kHz
d) 25kHz

26. A high first i.f allows easier filtering to prevent:
a) power supply ripple
b) local oscillator breakthrough
c) second channel interference
d) second i.f breakthrough

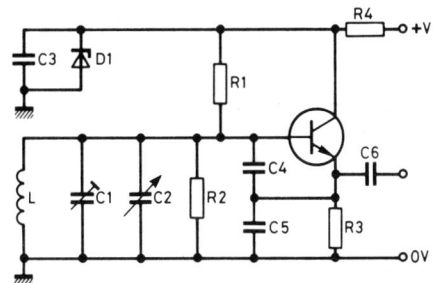

27. The above circuit is representative of:
a) an rf preamplifier
b) a crystal oscillator
c) a mixer
d) a vfo

28. In the circuit of Q27, which capacitor is the main tuning one?
a) C4
b) C3
c) C2
d) C1

29. In the circuit for Q27, why is the Zener diode used on the power supply?
    a) for fine tuning
    b) to help frequency stability
    c) for frequency modulation
    d) to clip a.m signals

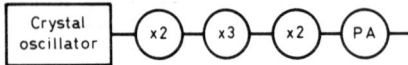

30. In the arrangement above, to give an output of 145.000MHz the crystal oscillator frequency must be:
    a) 10.7142MHz
    b) 12.08333MHz
    c) 72.500MHz
    d) 145.000MHz

31. Over driving a power amplifier will:
    a) give a high swr
    b) give minimum distortion on receive
    c) generate excessive harmonics
    d) minimise power output

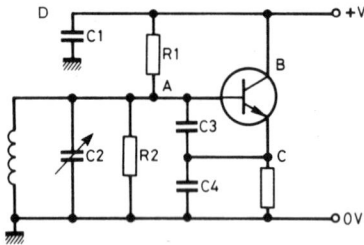

32. The above circuit is that of:
    a) a Colpitt's oscillator
    b) an rf preamplifier
    c) a crystal oscillator
    d) a tunable filter

33. In the circuit of Q32, the main tuning capacitor is:
    a) C1
    b) C2
    c) C3
    d) C4

34. Again referring to the circuit of Q32, the output is taken from:
    a) A
    b) B
    c) C
    d) D

35. Which of the following class of amplifier would be most efficient at amplifying an fm signal?
    a) Class A
    b) Class AB
    c) Class B
    d) Class C

36. The coil forming part of the frequency determining elements in a vfo should be:
    a) wound on a steel former
    b) made of resistance wire
    c) self supporting
    d) of sound mechanical construction

37. The output amplifier of an ssb transmitter must:
    a) act as a switch
    b) be in a linear mode
    c) be in a non-linear mode
    d) act as a multiplier

38. The units of the E field are:
    a) ohms
    b) volts
    c) tesla
    d) volts/metre

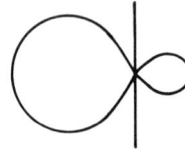

39. The above radiation pattern is typical of:
    a) a half wave dipole
    b) a quarter wave antenna
    c) a beam antenna
    d) a Marconi antenna

40. The velocity of propagation in free space is:
    a) 2×10^8 m/sec
    b) 2.5×10^8 m/sec
    c) 3×10^8 m/sec
    d) 3.5×10^8 m/sec

41. In daylight hours, which of the bands below would be suitable in working between Lands End and John O' Groats?
    a) 160m
    b) 80m
    c) 40m
    d) 15m

42. A sunspot cycle lasts about:
    a) 3 years
    b) 5 years
    c) 11 years
    d) 20 years

43. The ionospheric layer mainly responsible for long distance communication at hf is:
    a) D
    b) E
    c) F2
    d) F1

44. If two signals arrive at a point out of phase then:
    a) fading will occur
    b) signal enhancement occurs
    c) cross polarisation is produced
    d) the antenna impedance varies

45. A vertical antenna will provide:
    a) circular polarisation
    b) high angle radiation
    c) low angle radiation
    d) elliptical polarisation

46. The above represents a trap dipole for 3 bands: 20, 15 and 10 metres. The traps marked X will tune to:
    a) 12.4MHz
    b) 14.2MHz
    c) 21.2MHz
    d) 29MHz

47. An ammeter is to be used to measure rf current in a dummy load. The type should be:
    a) a moving coil meter
    b) a moving iron meter calibrated to 50Hz
    c) a thermocouple instrument
    d) none of these

48. The ammeter used in Q47 reads 2A when placed in series with a 50 ohm dummy load. The power in the load is:
    a) 25W
    b) 100W
    c) 200W
    d) 5000W

49. The typical accuracy of a moving coil meter is:
    a) 0.03%
    b) 0.3%
    c) 3%
    d) 10%

50. The current that must be taken for full scale deflection in a meter quoted as 10 kohm/V is:
    a) $10\mu A$
    b) $50\mu A$
    c) $100\mu A$
    d) $200\mu A$

51. The so called standing wave ratio meter physically measures:
    a) the forward and reverse impedances
    b) the actual forward and reverse power
    c) the forward and reverse voltages
    d) the cable characteristic impedance

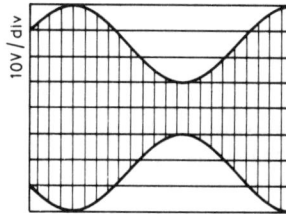

52. The above trace represents that from an oscilloscope monitoring an a.m transmission. The depth of modulation is:
    a) 40%
    b) 50%
    c) 60%
    d) 70%

53. The resonant frequency of a tuned circuit can be checked by:
    a) a dc voltmeter
    b) a dip oscillator
    c) a digital frequency meter
    d) an ohm-meter

54. A thermocouple instrument responds to:
    a) resistance
    b) temperature difference
    c) electric field
    d) magnetic field

55. Peak envelope power is defined as:
    a) the average power of an ssb transmission
    b) the average power at the peak of the envelope
    c) the peak to peak power at the crest of the modulation envelope
    d) the minimum power at the peak of modulation

B

Sample examination 2, Paper 1
Licensing conditions, transmitter interference and electro-magnetic compatibility.

1. The initial period for an amateur radio licence is:
    a) one year
    b) six months
    c) five years
    d) in perpetuity

2. The prefix to be used in Guernsey is:
    a) GJ
    b) GG
    c) GC
    d) GU

3. Data transmissions can be used:
    a) on all amateur bands
    b) on vhf only
    c) on uhf only
    d) on certain specified amateur bands

4. Calls must not be broadcast to amateur stations in general unless:
    a) it is to broadcast amateur news
    b) it is a CQ call
    c) it is to give a weather report
    d) there is a lift on

5. Entries in the log may not be:
    a) typed
    b) in indelible pencil
    c) in ink
    d) in pencil

6. As well as amateur frequency transmissions the licence allows reception of:
    a) diplomatic messages
    b) standard frequency transmissions
    c) news agency transmissions
    d) police transmissions

7. Messages containing which of the following is expressly forbidden:
    a) ASCII
    b) International No 2 Code
    c) Secret cypher
    d) Baudot Code

8. The suffix when riding a bicycle is:
    a) /B
    b) /P
    c) /M
    d) /C

9. Frequency modulation using voice is termed:
    a) F1A
    b) F3C
    c) F2A
    d) F3E

10. The maximum carrier power permitted in the 1.81 - 2.0MHz band is:
    a) 9dBW
    b) 16dBW
    c) 20dBW
    d) 26dBW

11. A class B licence allows operation above:
    a) 30MHz
    b) 50MHz
    c) 70MHz
    d) 144MHz

12. The licence conditions require that the purity of the transmitter output should be checked:
    a) weekly
    b) from time to time
    c) daily
    d) never

13. The amateur station can be inspected:
    a) only in office hours
    b) Saturday and Sunday only
    c) at any reasonable time
    d) between 1800 and 2100 on weekdays

14. In which of the following bands is it permissible to use fast scan tv?
    a) 28-29.7MHz
    b) 70.025-70.5MHz
    c) 144-146MHz
    d) 432-440MHz

15. Providing prior written notice has been given to the appropriate authority for operation from a different address in the UK, the suffix to be used is:
    a) /A
    b) /T
    c) /P
    d) not required

16. The calibration of a vfo in a transmitter varies by 0.05% over a given temperature range. How close can one go to the 29.7MHz setting in order to ensure staying within band at all temperatures within the stated range?
    a) 5kHz
    b) 15kHz
    c) 50kHz
    d) 150kHz

17. Key clicks from a cw transmitter can be reduced by:
    a) using screened lead to the key
    b) using a key click filter
    c) using a very small gap on the key
    d) a very controlled movement of the key

18. To generate an accurate and stable frequency, a frequency synthesiser shall:
    a) have a crystal reference oscillator

b) have no oscillator
c) run from a modulated dc supply
d) use the mains as a reference frequency

19. Which of the following arrangements will give the greatest accuracy when trying to check the carrier frequency of an fm transmitter?
a) an oscilloscope and an unmodulated carrier
b) digital frequency counter and a modulated carrier
c) digital frequency counter and an unmodulated carrier
d) absorption wavemeter and an unmodulated carrier

20. If over-deviation occurs in an fm transmitter this causes:
a) no sidebands
b) only a single sideband
c) several sets of sidebands
d) none of these

21. Which of the following antenna arrangements is least likely to radiate harmonics:
a) a dipole fed with coaxial cable
b) a dipole fed with balanced feeder
c) an inverted L-Marconi with vertical feeder
d) a trap dipole

22. Which of the plots below represents a filter suitable for following a microphone in order to limit the bandwidth?

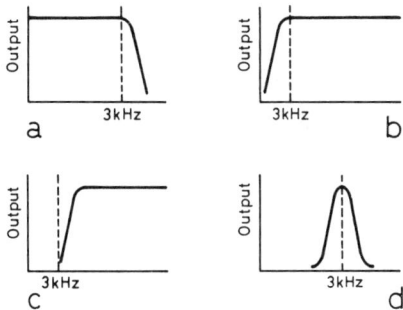

23. In order to minimise unwanted radiation, a mixer stage should be:
a) af decoupled
b) well screened
c) not earthed
d) supplied with mains voltage

24. Parasitic oscillations are caused by:
a) mains hum
b) mains ripple
c) unwanted semiconductors
d) self resonance of parts of an amplifier circuit

25. The equipment at an amateur station should be so designed, constructed or maintained so that:
a) it does not cause undue interference with any wireless telegraphy
b) it causes interference with any wireless telegraphy
c) it will transmit automatically
d) it operates outside the specified bands

26. To minimise rf interference it is wise to:
a) bond everything to a mains water pipe
b) use separate earthing points for equipment
c) use a separate rf return
d) not earth equipment

27. Earth return circuits should always be:
a) high impedance
b) highly reactive
c) low impedance
d) inductive

28. An external antenna is always preferred because:
a) it radiates less harmonics
b) it will be subject to less radiation
c) the coupling to mains wiring is minimised
d) the transmitted signal is less

29. A 12MHz crystal oscillator is followed by several multiplier stages. The first three are ×2×2×2. This might cause interference on:
a) medium wave broadcasts
b) fm broadcast band 88 - 108MHz
c) uhf tv band
d) none of these

30. Power supplies to rf power amplifiers should:
a) be open wires
b) be af filtered
c) be rf filtered
d) be inductively coupled

31. Instead of a braid breaking choke to stop tvi it might be possible to use:
a) resistors
b) a mains auto transformer
c) an rf isolation transformer
d) no screening

32. The use of indoor transmitting antennas:
a) should be encouraged
b) will never couple with the mains
c) have a good chance of coupling into the mains
d) give more long distance contacts

33. The type of transmission most prone to causing interference to an audio amplifier system is:
a) frequency modulation
b) frequency shift keying
c) amplitude modulation
d) phase modulation

34. A 435MHz high gain antenna points straight into a uhf tv receiving antenna. This could cause:
    a) problems with the 435MHz receiver
    b) overloading of the tv front end
    c) self oscillation of the 435MHz transmitter
    d) melting of the tv antenna elements

35. The medium waveband is prone to:
    a) second channel problems from uhf transmitters
    b) second channel problems from vhf transmitters
    c) second channel problems from 28MHz transmitters
    d) second channel problems from 1.8MHz transmitters

36. A corroded connector on a neighbour's tv receiving antenna may cause:
    a) unwanted mixing products due to it exhibiting diode properties
    b) mains rectification
    c) enhanced signal reception due to its filtering properties
    d) increased amplification

37. A transmitter is connected by a short coaxial cable to a colinear antenna with 6dB gain. When the output power to the antenna is reduced to 5W no more interference is caused to a neighbour's hi-fi system. This corresponds to an effective radiated power of:
    a) 1W
    b) 10W
    c) 11W
    d) 20W

38. In an attempt to get rid of rf interference in an audio power amplifier a disc ceramic capacitor could be fitted:
    a) across the base-emitter junction of the audio power transistor
    b) across the base-collector junction of the audio power transistor
    c) between ground and emitter
    d) between the collector and emitter leads of the audio power transistor

39. If a neighbour complains of "breakthrough" one's immediate response should be:
    a) blame the neighbour's equipment straightaway
    b) tell them you will do nothing about it
    c) keep polite and get them to help you investigate
    d) inform RSGB and DTI immediately

40. Which of the following sets of components are used to make rf filters?
    a) diodes and resistors
    b) Zener diodes and inductors
    c) l.e.ds and capacitors
    d) inductors and capacitors

41. The insertion loss in the passband of a passive high pass filter for a tv downlead should be in the range:
    a) –6 to 0dB
    b) 0 to 6dB
    c) 12 to 18dB
    d) 24 to 30dB

42. The impedance a filter should match for use in a uhf tv downlead is normally:
    a) 0 ohms
    b) 25 ohms
    c) 50 ohms
    d) 75 ohms

43. When operating at hf, interference is caused on a tv. The most likely route for the interfering signal is:
    a) via the earth
    b) through the transmitter power supply
    c) by the tv antenna coaxial cable screen and/or i.f stages
    d) by frequency multiplication in free space

44. When living in a row of terraced houses and to minimise the possibility of interference the best route for an hf wire antenna is:
    a) from a joint chimney stack and round the tv antenna
    b) along the row of houses at gutter height
    c) at right angles to the row of houses and going away from them
    d) within the roof space of the row of houses

45. The second harmonic of a 435MHz transmission lies:
    a) in a police band
    b) between 1GHz and 10GHz
    c) in a uhf tv band
    d) below 144MHz

Sample examination 2, Paper 2
Operating practices, procedures and theory.

1. Which of the following uses the International Phonetic Alphabet?
    a) alpha, norway, delta
    b) charlie, alpha, tosca
    c) denmark, oscar, gordon
    d) bravo, uniform, golf

2. The Q code used by amateurs to mean change frequency is:
    a) QSB
    b) QRG
    c) QRF
    d) QSY

3. For safety reasons, across high value capacitors there should always be:
    a) an inductor
    b) a bleed resistor
    c) a short circuit
    d) an open circuit

4. When replying to a CQ call on telegraphy:
    a) reply at a speed you can receive
    b) reply at a faster cw speed
    c) reply using telephony
    d) reply always with QRZ

5. The band plans in the UK are:
    a) to be observed for good operating practice
    b) applicable only to contacts outside the UK
    c) mandatory
    d) applicable only to inter-UK contacts

6. The tone required for repeater access is:
    a) 1725Hz
    b) 1750Hz
    c) 1775Hz
    d) 1800Hz

7. When making a CQ call it is good practice to:
    a) use a frequency occupied by a weak station
    b) always use cw
    c) only call dx stations
    d) check that the frequency is clear before starting

8. In the RST code, which of the following represents a perfectly readable signal?
    a) R1
    b) R5
    c) S5
    d) S9

9. In order to activate a UK repeater, it is necessary to:
    a) obtain a special licence
    b) be a member of the local repeater group
    c) use a toneburst
    d) use cw

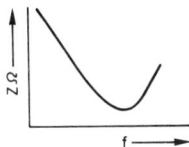

10. The above impedance-frequency curve represents:
    a) a capacitance
    b) a parallel tuned circuit
    c) an inductance
    d) a series tuned circuit

11. As the frequency rises the reactance of an inductor:
    a) stays constant
    b) decreases

c) increases
d) none of these

12. The prefix micro is equivalent to:
    a) 10^{-6}
    b) 10^{-3}
    c) 10^3
    d) 10^6

13. The peak value of the 240V mains is:
    a) 168V
    b) 339V
    c) 480V
    d) 678V

14. The resonant frequency of a tuned circuit is 1MHz and the bandwidth at the −3dB point is 10kHz. The Q factor is:
    a) 50
    b) 100
    c) 500
    d) 1000

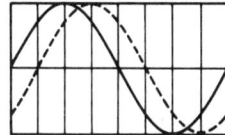

15. The phase difference between the two sine waves in the above diagram is:
    a) 0 deg
    b) 45 deg
    c) 90 deg
    d) 180 deg

16. The above symbol represents:
    a) a dual bipolar transistor
    b) a dual diode
    c) a dual varactor diode
    d) a dual gate mos fet

17. A forward biased pn junction allows:
    a) current to flow from p to n region
    b) current to flow from n to p region
    c) no current to flow
    d) electrons to flow from p to n region

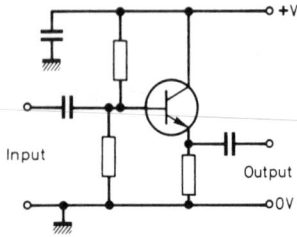

18. The above circuit is:
   a) a common emitter amplifier
   b) a buffer
   c) an oscillator
   d) a common base stage

19. The phase difference between input and output voltages of the circuit of Q18 is
   a) 0 deg
   b) 45 deg
   c) 90 deg
   d) 180 deg

20. The output impedance of the circuit shown in Q18 is:
   a) very high
   b) infinite
   c) very low
   d) inductive

21. One use of the circuit in Q18 would be:
   a) as a buffer between an oscillator and load
   b) as a high gain voltage amplifier
   c) as a voltage inverter
   d) as a frequency multiplier

22. An integrated circuit is:
   a) a passive device only
   b) a complete set of capacitors
   c) an encapsulated complex circuit
   d) a discrete component circuit

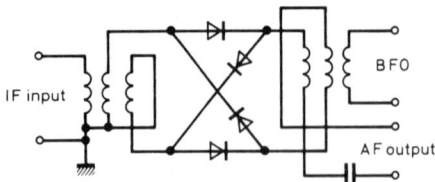

23. The above circuit is typical of:
   a) a mains rectifier
   b) a buffer circuit
   c) a voltage stabiliser
   d) a product detector

24. The output from the circuit in Q23 is:
   a) very low impedance
   b) high current

c) the difference between the two input frequencies
d) the sum of the two input frequencies

25. A ratio detector is normally used to demodulate:
   a) a.m signals
   b) dsb signals
   c) fm signals
   d) ssb signals

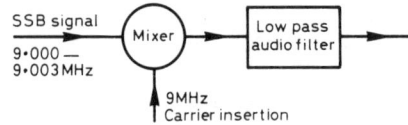

26. In the mixing arrangement shown above, the output from the low pass filter will be in the range:
   a) 0 - 3kHz
   b) 0 - 3MHz
   c) 18.000 - 18.003MHz
   d) 81.000 - 81.027MHz

27. A notch filter is useful for:
   a) reducing the vfo frequency
   b) reducing a narrow-band interfering signal
   c) reducing the power supply voltage
   d) reducing the wanted signal

28. An active mixer will usually provide:
   a) conversion loss
   b) audio distortion
   c) conversion gain
   d) insensitive receivers

29. One advantage of fm over a.m is:
   a) increase range in distance
   b) narrower bandwidth required
   c) freedom from most sources of man made interference
   d) lower cost of equipment

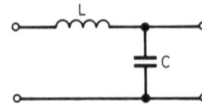

30. The above circuit immediately follows a microphone amplifier in a transmitter. Its purpose is:
   a) to increase high frequency content of the signal
   b) limit audio bandwidth
   c) limit amplitude of the audio signal
   d) provide rf compression

31. A cw transmitter has a power output of 100W. This is equivalent to:
   a) 10dBW
   b) 20dBW

c) 22dBW
d) 26dBW

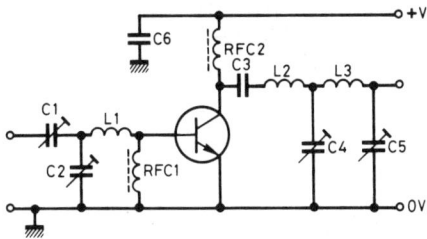

32. The above circuit is a power amplifier. It operates in:
a) Class A
b) Class AB
c) Class B
d) Class C

33. If the amplifier of Q32 is used after a transceiver it is only useful for amplifying:
a) a cw transmission
b) an ssb transmission
c) an ssb transmission with reduced carrier
d) an a.m transmission

34. In the circuit of Q32 L2,L3, C4 and C5 form:
a) a band pass filter
b) a high pass filter
c) a notch filter
d) a low pass filter

35. In the circuit of Q32 C1, C2 and L1 are for:
a) frequency stability
b) bias adjustment
c) gain adjustment
d) impedance matching

36. The power supply to a vfo should be:
a) well regulated
b) straight from the smoothing circuit
c) unregulated
d) ac only

37. The typical deviation of an amateur nbfm signal is:
a) 5kHz
b) 7.5kHz
c) 10kHz
d) 25kHz

38. If the far end of a coaxial cable becomes open circuited, the voltage there:
a) can rise to a high value
b) will always be zero
c) is the same as the matched value
d) is converted to dc

39. The major mode of propagation at vhf over long distances is known as:

a) tropospheric propagation
b) ionospheric propagation
c) ground wave propagation
d) additive propagation

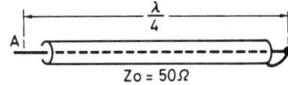

40. In the above diagram what is the impedance seen at end A?
a) zero
b) 50 ohms
c) 150 ohms
d) infinite

41. The dielectric of open wire feeder is usually:
a) air
b) polythene
c) rubber
d) water

42. The wavelength of a signal in free space with a frequency of 100MHz is:
a) 30mm
b) 0.3m
c) 3m
d) 30m

43. The beamwidth of an antenna is normally measured at the:
a) −1dB point
b) −3dB point
c) −6dB point
d) −9dB point

44. The side of a quad antenna is:
a) a quarter wavelength
b) a half wavelength
c) three quarters of a wavelength
d) a full wavelength

45. The polarisation of an electromagnetic wave is defined by the direction of:
a) the H field
b) propagation
c) the E field
d) the receiving antenna

46. Auroral reflections occur from which of the following areas?
a) polar regions
b) equatorial regions
c) lunar regions
d) tropical regions

47. The instrument that can be used to check the resonant frequency of a tuned circuit is:
a) an absorption wavemeter

b) a digital frequency counter
c) a dip oscillator
d) a multimeter

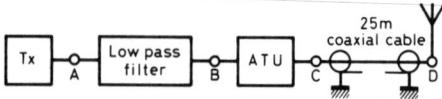

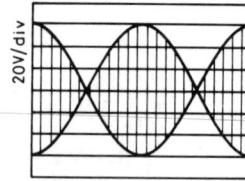

48. The above represents a typical arrangement of equipment. The swr meter should be placed at:
a) A
b) B
c) C
d) D

49. A moving coil meter by itself only responds to:
a) ac
b) dc
c) ac and dc
d) frequency

50. An swr meter is used to check:
a) transmitter efficiency
b) harmonic output
c) transmitter bandwidth
d) transmitter to antenna matching

51. If the connecting cable just after an swr meter in an hf system goes open circuit the vswr will be:
a) 1:1
b) very low
c) reasonably low
d) very high

52. The above trace is typical of:
a) an fm waveform
b) a two tone ssb test
c) a 100% a.m signal
d) a cw signal

53. The trace shown in Q52 is taken across a 50 ohm resistive load. What is the average power at the peak of the envelope?
a) 3.6W
b) 7.2W
c) 36W
d) 72W

54. If a divide by ten prescaler is placed ahead of a frequency counter and the value shown is 43.3350MHz, the true value is:
a) 0.433350MHz
b) 4.33350MHz
c) 433.350MHz
d) 4333.50MHz

55. The horizontal, X axis, on an oscilloscope displays:
a) voltage
b) capacitance
c) reactance
d) time

Sample examination 3, Paper 1

Licensing conditions, transmitter interference and electro-magnetic compatibility.

1. The Amateur Licence can be revoked by a broadcast by:
   a) BBC
   b) IBA
   c) RSGB
   d) another amateur

2. The lower age limit for passing the Radio Amateurs' Examination is:
   a) not specified
   b) 12 years of age
   c) 13 years of age
   d) 14 years of age

3. Details of transmissions made when operating as / M:
   a) must be entered in the main log
   b) must be entered in the mobile log
   c) need not be recorded
   d) must be tape recorded

4. Double sideband, suppressed carrier transmissions are:
   a) prohibited
   b) permitted
   c) allowed with written permission
   d) allowed only on vhf frequencies

5. Which of the following bands is used by amateurs in the UK on a secondary basis?
   a) 1.81 - 2.0MHz
   b) 10.1 -10.15MHz
   c) 21.0 - 21.45MHz
   d) 144 -146MHz

6. Which of the following types of message can be sent?
   a) those of a personal nature
   b) those of an obscene nature
   c) those of a misleading nature
   d) those of a religious nature

7. G7xxx when on holiday in the Shetland Islands should sign:
   a) G7xxx/GM
   b) GS7xxx
   c) GM/G7xxx
   d) GM7xxx

8. Items that can be inspected by a person acting under the authority of the Secretary of State are:
   a) the station only
   b) the log only
   c) the station and the log
   d) the station, log and licence

9. The licence requires that the receiver(s) at an amateur station must be capable of receiving:

   a) only fm transmissions
   b) one mode of emission
   c) ssb transmissions only
   d) any mode on which the station can transmit

10. Using the standard codes for modulation, the designation F3E stands for:
    a) telegraphy by frequency shift keying
    b) telegraphy by on-off keying of a modulating audio frequency
    c) frequency modulation by voice
    d) facsimile transmission

11. An Amateur Licence A requires which of the following?
    a) a pass in the RAE and the morse test
    b) a pass in the RAE only
    c) a pass in the morse test only
    d) to have held a class B licence previously

12. In the event of a national disaster, which of the following bands can be used by non-amateurs in the disaster area in accordance with the licensing conditions?
    a) 144 - 146MHz
    b) 29.0 - 29.7MHz
    c) 28.0 - 29.0MHz
    d) none of these

13. An amateur has a tower 13m high on which he wants to place a vertical antenna: he lives within 0.5km of the boundary of an airfield. The maximum height antenna he can use is:
    a) 0.5m
    b) 1.25m
    c) 2m
    d) 3m

14. Fast scan tv transmission is permitted on:
    a) 14MHz band
    b) 21MHz band
    c) 144MHz band
    d) 10GHz band

15. The amateur licence also allows transmissions on behalf of:
    a) any third party
    b) private taxi firms
    c) a public transport concern
    d) County Emergency Planning Officer

16. The bandwidth of a data transmission should be kept to that of telephony in order to:
    a) help demodulation
    b) conserve bandwidth
    c) reduce transmitter power
    d) reduce self oscillation

17. If the coil in a vfo has no former, then vibrations:
    a) will keep the frequency generated in the band
    b) may take the frequency generated out of band

c) will provide a clean signal
d) are beneficial

18. Which of the following would be useful in rejecting an unwanted signal at the input to a receiver?

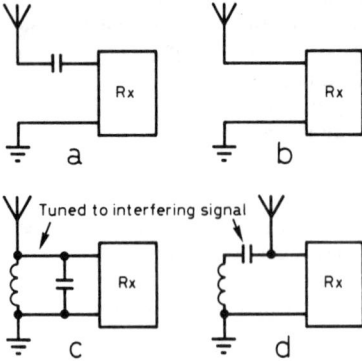

19. The possibility of harmonic radiation should be checked with an absorption wavemeter:
a) daily
b) only when a complaint arises
c) on receiving an interfering signal
d) from time to time

20. When transmitting cw a modulation envelope as shown below should be avoided because:

a) it would be difficult to read
b) of problems in the p.a during the spaces
c) key clicks might be produced
d) the power supply would become unstable

21. If the accuracy of a digital readout on a transmitter is 10 parts per million, a frequency shown of 14.25MHz could be as high as:
a) 14.250001425MHz
b) 14.25001425MHz
c) 14.2501425MHz
d) 14.151425MHz

22. In order to minimise the risk of self-oscillation in rf circuits:
a) each stage should be well screened
b) screening should not be used between stages
c) stages should be connected with open wires
d) every other stage should be screened

23. In order to restrict the range of audio frequencies from a microphone, it should be followed by:
a) a high pass filter
b) a resistive divider

c) a low pass filter
d) an amplitude limiter

24. If the vfo of a transmitter is subject to varying temperatures, this might cause:
a) chirp
b) drift
c) harmonic generation
d) no problems

25. An overdeviated fm transmission will produce:
a) excessive sidebands
b) one sideband
c) only two sidebands
d) none of these

26. Stages where harmonic generation occurs should be:
a) sealed in epoxy resin
b) open to the atmosphere
c) encased in polystyrene
d) screened very carefully

27. Spurious resonances may occur in decoupling circuits due to:
a) the power supply
b) self resonance of rf chokes
c) saturation of the core of rf chokes
d) the resistive element of an rf choke

28. A multi-band antenna is:
a) less likely to radiate harmonics
b) more likely to radiate harmonics
c) never going to radiate harmonics
d) more efficient than a dipole

29. Wideband frequency modulation in a nbfm section of a band:
a) will cause no problems
b) will cause adjacent channel interference
c) helps harmonic rejection
d) does not cause parasitic oscillations

30. A transmitter is using upper sideband and set to 14.349MHz. It is modulated by audio with up to 3kHz. This will:
a) be within the allotted allocation
b) go above the band edge
c) have zero sidebands
d) cause over deviation

31. A neighbour using a tv set top antenna complains of interference when you are transmitting at vhf. As a first step to eliminating this problem you could suggest:
a) better coaxial cable on their antenna
b) they use a preamplifier
c) they use a roof mounted antenna
d) their set is no good

32. In making decisions on type of feeder for a transmitting antenna, then to minimise the risk of interference:

a) consider only long wire feeds
b) use unscreened feeders near the building
c) use only screened feeders near the building
d) do not earth any part of the feeders

33. When operating a mobile hf set at home from a battery supply and using the base antenna there is no interference problem. When using the same arrangement but with an earthed battery charger connected interference occurs on an electronic organ. The possible cause is:
a) the production of sub-harmonics at the transmitter
b) very strong received signals
c) poor rf earthing
d) that the rf earthing is too good

34. A 432MHz amateur station causes interference to a nearby tv receiver. Which of the following filters could be fitted in the tv download in order to minimise the interference problem?

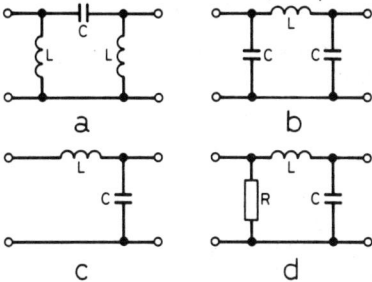

35. A transmitter/antenna system has an erp of 225W, at what distance is the field strength 0.5V/m?
a) 0.21m
b) 2.1m
c) 21m
d) 210m

36. In the diagram above the former on which the cable is wound to make a mains filter choke is likely to be:
a) plastic
b) steel
c) ferrite
d) paramagnetic

37. When operating on any band interference is experienced on both yours and your neighbours telephones. These are both the same model. Which is likely to be the most probable cause?
a) induction via the mains wiring
b) the telephone instruments and wiring
c) the transmitter local oscillator
d) the telephone exchange

38. The type of interference caused by a transmitter can be classified as:
a) broad band
b) pseudo random
c) white noise
d) narrow band

39. If the mains earth is used as an rf earth, this could be prone to causing:
a) mains hum
b) parasitic oscillations
c) mains borne interference
d) self oscillation

40. How far should a tv antenna be placed from a transmitting antenna in order to reduce the risk of interference?
a) as far as possible
b) as close as possible
c) 1.098 metres
d) half a wavelength away at 432MHz

41. A tv still under guarantee suffers interference from a nearby transmitter. The transmitter output is proved to be "clean" and the provision of a filter in the tv down lead has little effect. A sensible approach might be:
a) tell the tv user not to use the tv
b) tell the tv user to return the set to the shop and tell them it lacks immunity and demand it be put right
c) send the tv set the DTI
d) get the tv user to fit filters in the transmitter output lead

42. In making a filter to minimise breakthrough, when cutting the coaxial cable to solder to the connecting socket, one should:
a) cut as much screening braid off as possible
b) short the screening braid to the centre conductor
c) cut as little screening braid off as possible
d) earth the centre conductor

43. The insertion loss of a high pass filter in a tv download is measured in:
a) decibels
b) watts
c) Ohms
d) negative Gerbils

44. If a strong rf signal is proved to be induced into the i.f stage of a neighbour's broadcast receiver, this can be referred to as:
    a) intermediate pick up
    b) frequency pick up
    c) Faraday pick up
    d) direct pick up

45. If after extensive attempts at reducing the interference caused on a tv by a nearby amateur transmitter fails, one could:
    a) get the RSGB to look at the tv set
    b) get the Radio Investigation Service to investigate further
    c) send the transmitter to the DTI
    d) none of these

Sample examination 3, Paper 2
Operating practices, procedures and theory.

1. The Q code for 'stand by' is:
    a) QRN
    b) QRM
    c) QRS
    d) QRX

2. In a telephony contact it is advisable to:
    a) speak as fast as possible in order to clear the frequency
    b) speak very slowly using phonetics as often as possible
    c) use Q codes as often as possible
    d) speak clearly and not too quickly

3. The recommended phonetic spelling of VALVE is:
    a) volt, alpha, lima, volt, echo
    b) victor, alpha, london, volt, echo
    c) victor, america, lima, victor, echo
    d) victor, alpha, lima, victor, echo

4. It is not good practice to:
    a) use double insulated cable on eht circuits
    b) use a separate rf earth
    c) use a gas pipe for the earth connection
    d) have safety switches

5. The only general call allowed from an amateur station is:
    a) a news bulletin
    b) a CQ call
    c) a third party call
    d) on vhf

6. Directional CQ calls should:
    a) not be made
    b) be ignored
    c) only be by cw
    d) be respected

7. When working through a satellite:
    a) only use sufficient power to maintain reliable communication
    b) it is necessary to have special permission from the Licensing Authority
    c) one must be a member of AMSAT
    d) high power must be used

8. In the UK, a 2m repeater is accessed by a tone of:
    a) 1650Hz
    b) 1700Hz
    c) 1750Hz
    d) 1800Hz

9. Having established contact with another station on a calling frequency, one should:
    a) continue the QSO on that frequency
    b) move to another frequency, having checked that it is clear
    c) move to an agreed frequency, even if it is already in use
    d) remain on the calling frequency and invite other stations to join the QSO

10. What is 0.01ms expressed in μs?
    a) 0.00001μs
    b) 0.1μs
    c) 1μs
    d) 10μs

11. The tolerance of a resistor is given as 10%. If the resistor value is 4.7 kohm, its value must lie between:
    a) 4230 and 5170 ohms
    b) 4653 and 4747 ohms
    c) 4230 and 4747 ohms
    d) 4653 and 5170 ohms

12. 0.1V expressed as mV is:
    a) 1mV
    b) 10mV
    c) 100mV
    d) 1000mV

13. Which of the following does not rely on a magnetic field:
    a) a dynamic microphone
    b) a loudspeaker
    c) a carbon microphone
    d) a transformer

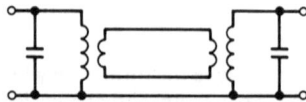

14. The above circuit is coupled:
    a) electrostatically
    b) capacitively
    c) magnetically
    d) resistively

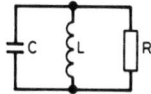

15. The above circuit is that of:
   a) a series tuned circuit
   b) a low pass filter
   c) a high pass filter
   d) a parallel resonant circuit

16. N type material has:
   a) a deficiency of electrons
   b) additional holes
   c) a deficiency of an atom
   d) an excess of electrons

17. The above symbol represents:
   a) a pnp transistor
   b) an npn transistor
   c) a p-type fet
   d) an n-type fet

18. The main purpose of a Varactor diode is:
   a) tuning
   b) rectification
   c) voltage regulation
   d) display

19. In a Class B amplifier, using an npn transistor, the base bias potential is:
   a) much greater than the emitter potential
   b) the same as the collector potential
   c) about 0.6V above the emitter voltage
   d) less than the emitter voltage

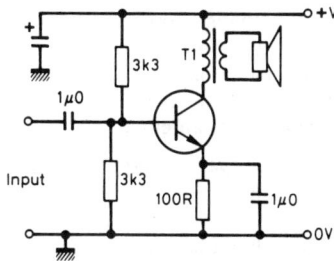

20. The above circuit is that of:
   a) an audio amplifier
   b) an rf amplifier
   c) a mixer
   d) a bfo

21. The purpose of transformer T1 in Q20 is:
   a) voltage amplification
   b) impedance matching
   c) power amplification
   d) gain control

22. The circuit in Q20 would operate in:
   a) Class A
   b) Class AB
   c) Class B
   d) Class C

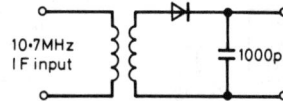

23. The above circuit is:
   a) a power rectifying circuit
   b) an fm discriminator
   c) a varactor tuner
   d) an a.m detector

24. The typical first i.f in a vhf receiver is:
   a) 3kHz
   b) 50kHz
   c) 455kHz
   d) 10.7MHz

25. For reception of cw signals, the beat note produced by the bfo should be about:
   a) 400Hz
   b) 800Hz
   c) 2kHz
   d) 456kHz

26. If the value of the capacitor in a receiver vfo increases with temperature, then for increased temperature the vfo frequency will:
   a) stay the same
   b) go to zero
   c) increase
   d) decrease

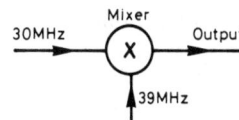

27. The principle outputs from the above circuit are:
   a) 9 and 39MHz
   b) 9 and 69MHz
   c) 30 and 39MHz
   d) 39 and 69MHz

28. If the arrangement of Q27 is followed by a 9MHz filter, what frequency signal might cause second

channel interference?
a) 18MHz
b) 21MHz
c) 39MHz
d) 48MHz

29. The sensitivity of a receiver specifies:
a) the bandwidth of the rf preamplifier
b) the stability of the vfo
c) its ability to receive weak signals
d) its ability to reject strong signals

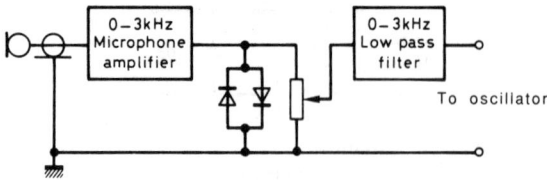

30. The above circuit is typical of:
a) rf clipping
b) deviation control
c) amplitude modulation
d) ssb generation

31. If the power supply to an output stage is modulated, this produces:
a) a.m
b) fm
c) nbfm
d) pm

32. A crystal oscillator has inherently:
a) poor frequency stability
b) fm components
c) good frequency stability
d) vibration problems

33. The output of a balanced mixer has:
a) the full carrier present
b) many mixing products present
c) a reduced carrier plus sidebands
d) only the two sidebands

34. An amplifier for ssb use only:
a) operates in Class C
b) must cope with a 100% duty cycle
c) requires no biasing
d) need not be rated for a 100% duty cycle

35. Varying the capacitance of a Varactor diode connected across a crystal oscillator is one method of producing:
a) amplitude modulation
b) double sideband modulation
c) cw signals
d) frequency modulation

36. The L-C circuits immediately following a power

amplifier allow:
a) harmonic analysis
b) harmonic amplification
c) impedance mismatching
d) impedance matching

37. An 8MHz oscillator is used to generate a 144MHz cw signal. The multiplication is:
a) ×3×3×3
b) ×3×3×2
c) ×3×2×2
d) ×2×2×2

38. Which is the lowest ionospheric region?
a) F1
b) F2
c) D
d) E

39. The half wavelength in free space at 30MHz is:
a) 0.5m
b) 5m
c) 10m
d) 20m

40. The relationship between velocity of propagation, frequency and wavelength is:
a) $v = f \times \lambda$
b) $\lambda = f \times v$
c) $f = v \times \lambda$
d) $v \times f \times \lambda = 1$

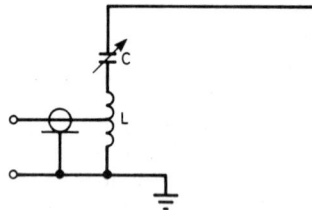

41. The above antenna is known as:
a) a Marconi type antenna
b) a Yagi type antenna
c) a Cubical Quad antenna
d) a W3DZZ dipole antenna

42. In the diagram of Q41, L and C are for:
a) the earth connection
b) increasing the radiation resistance
c) decreasing the radiation resistance
d) matching the antenna to the transmitter

43. The signal returned from the layers above the earth are referred to as:
a) the ground wave
b) the ionospheric wave
c) the tropospheric wave
d) the direct wave

44. The order of the ionospheric layers starting at the earth's surface is:
    a) D,E,F2,F1
    b) D,E,F1,F2
    c) E,D,F1,F2
    d) F1,D,E,F2

45. At a sunspot minimum the normal long distance band is:
    a) 3.5MHz
    b) 7MHz
    c) 14MHz
    d) 28MHz

46. What is the frequency corresponding to a wavelength of 300mm in a vacuum?
    a) 100kHz
    b) 1MHz
    c) 100MHz
    d) 1GHz

47. Which of the following components could be attached to a moving coil meter in an attempt to measure power?
    a) resistor
    b) thermistor
    c) thermocouple
    d) therm

48. A good dummy load for rf is constructed from:
    a) light bulbs
    b) a column of water
    c) wirewound resistors
    d) non-reactive resistors

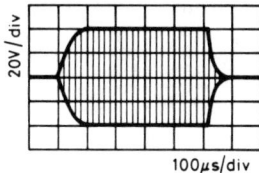

100μs/div

49. The above represents the voltage across a 50 ohm load during part of a cw transmission. This corresponds to an output power of:
    a) 0.4W
    b) 4W
    c) 8W
    d) 16W

50. The duration of the pulse shown in Q49 is:
    a) 5μs
    b) 50μs
    c) 350μs
    d) 500μs

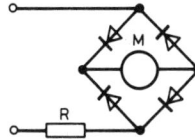

51. A moving coil meter arranged in the above manner allows measurement of:
    a) ac and dc
    b) ac only
    c) dc only
    d) frequency

52. Which of the following is most suitable for an accurate frequency measurement?
    a) an absorption wavemeter
    b) an oscilloscope
    c) a multimeter
    d) a digital frequency counter

53. The movement in a typical multimeter is:
    a) moving permanent magnet
    b) electrostatic
    c) non-magnetic
    d) moving coil

54. The control used on an oscilloscope to obtain a stable display is:
    a) the Y sensitivity
    b) the X sensitivity
    c) the trigger
    d) the calibrator

55. When monitoring the frequency of an unmodulated carrier, the readout of a digital frequency counter should show:
    a) the carrier frequency plus the number of significant harmonics present
    b) how many sidebands there are
    c) nothing
    d) the constant carrier frequency

Sample examination 4, Paper 1
Licensing conditions, transmitter interference and electro-magnetic compatibility.

1. A station can be closed down at any time by a demand from a person acting under the authority of:
   a) the local council
   b) the planning authority
   c) the licensing authority
   d) the local MP

2. The licensee must also comply with the relevant provisions of:
   a) the RSGB
   b) the FCC
   c) the local radio club
   d) the Telecommunications Convention

3. When in communication with another station, the callsign should be sent for identification purpose at least:
   a) every 5 minutes
   b) every 15 minutes
   c) half hourly
   d) hourly

4. The callsign of a station operating from the main address in Scotland must use the prefix:
   a) GM
   b) GS
   c) G
   d) Gxxxx/GM

5. What is the peak envelope power allowed on the 432 - 440MHz band?
   a) 15dBW
   b) 26dBW
   c) 28dBW
   d) 40dBW

6. In which of the following bands can slow scan television not be used ?
   a) 3.5 -3.8MHz
   b) 18.068 - 18.168MHz
   c) 28 - 29.7MHz
   d) 144 - 146MHz

7. H3E is the designation for:
   a) frequency modulation
   b) ssb with full carrier
   c) ssb with reduced carrier
   d) ssb with no carrier

8. Which of the following bands is not for amateur use only?
   a) 3.5 -3.8MHz
   b) 7.0 -7.1MHz
   c) 14.0 - 14.35MHz
   d) 21.0 - 21.45MHz

9. The emissions to be used by an amateur are:
   a) J3E and F3E only
   b) not specified
   c) those specified in the licence schedule
   d) as dictated by the BBC

10. The licence conditions state that:
    a) the licence is not transferable
    b) the licence is transferable
    c) the licence can never be revoked
    d) the licence gives a waiver over copyright

11. It is an offence to send by wireless telegraphy:
    a) certain misleading messages
    b) severe weather warnings
    c) test transmissions
    d) ASCII code

12. Two stations are in morse telegraphy contact and never drop below 25 wpm at any time. This is:
    a) within the terms of the licence
    b) within the terms of the licence except for callsign identification
    c) above the maximum speed permitted by the licence
    d) under the speed dictated by the licence

13. If G1xxx is on holiday in Scotland and operating hand held equipment from the top of Ben Nevis, the callsign used is:
    a) G1xxx/GM/P
    b) G1xxx/M
    c) GM1xxx/P
    d) GM1xxx/M

14. Which of the following is not acceptable for writing an entry in a log book?
    a) typewriter
    b) pencil
    c) ink
    d) dot matrix printer

15. In which part of the 430 - 440MHz band is power limited to 10dBW erp?
    a) 430 - 432MHz
    b) 432 - 434MHz
    c) 434 - 437MHz
    d) 437 - 440MHz

16. A low pass filter will:
    a) stop sub-harmonics
    b) suppress harmonics
    c) always eliminate interference
    d) improve harmonic radiation

17. The mains transformer in a transmitter is fitted with an internal screen. To minimise the possibility of introducing mains-borne interference it should be connected to:
    a) the chassis

b)   the vfo output
c)   the live side of the mains
d)   left floating

18.   Which of the following represents over-modulation of an a.m signal?

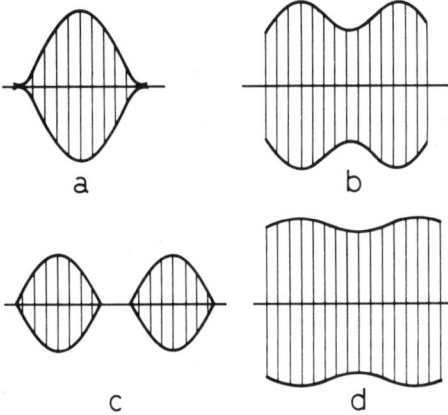

a   b

c   d

19.   Overmodulation in an a.m signal is likely to cause:
a)   excessive deviation
b)   10 sidebands
c)   minimum interference
d)   severe splatter

20.   If the mains earth is used as an rf earth, this could be prone to causing:
a)   mains borne interference
b)   mains hum
c)   parasitic oscillations
d)   self oscillation

21.   To stop rf going back into the mains cable, which of the following could be fitted in series with the mains lead?

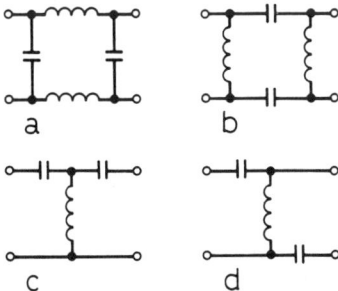

a   b

c   d

22.   Any device that exhibits diode properties may:
a)   reduce interference
b)   produce unwanted mixing products
c)   enhance the wanted signal
d)   reduce harmonic radiation

23.   If the power output meter of a Class C amplifier has a small varying reading when it is not being keyed, one might suspect:
a)   unregulated dc
b)   parasitic oscillations
c)   the reception of a strong carrier
d)   that there are no problems

24.   When using voice, which of the following modes of transmission requires the least bandwidth?
a)   single sideband
b)   amplitude modulation
c)   frequency modulation
d)   phase modulation

25.   When transmitting at 3.73MHz, the second harmonic will be:
a)   1.865MHz
b)   7.46MHz
c)   10.19MHz
d)   none of these

26.   To keep a shack tidy, an amateur puts all his mains cables and rf feeders in a single wiring loom. This:
a)   would represent the ideal thing to do
b)   might cause unwanted 50Hz modulation
c)   might be prone to producing mains borne interference
d)   will never produce interference tendencies

27.   Sparks generated at the contacts of a morse key:
a)   will cause long distance interference
b)   will cause short range interference
c)   produce no interference effects
d)   provide a good clean signal

28.   Cross modulation is caused by:
a)   the wanted complex signal mixing with itself
b)   a harmonic mixing with itself
c)   overdriving an amplifier
d)   a strong unwanted signal mixing with the wanted signal

29.   A long wire antenna that uses a vertical feed line near a house can:
a)   never cause interference
b)   induce strong unwanted signals into tv coaxial down leads
c)   cure all interference problems
d)   produce parasitic oscillations

30.   Multiplier stages should be:
a)   encased in plastic
b)   surrounded by long wires
c)   well screened
d)   tuned for maximum radiation

31.   In considering the equipment and power levels in a densely populated neighbourhood, it might be advisable to:

a) keep the antennas as low as possible
b) site antennas as remotely as possible from neighbours
c) use maximum possible power and erp on 432MHz
d) always use long wire feeds

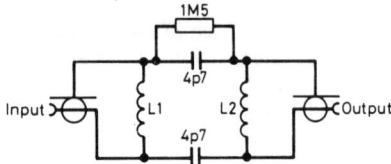

32. The above filter is:
   a) a band stop filter
   b) a notch filter
   c) a braid breaker and high pass filter
   d) a low pass filter only

33. A herring bone pattern is caused on a tv picture, this is likely to be caused by:
   a) poor commutation in an electric vacuum cleaner
   b) a nearby radio transmitter
   c) a facsimile receiver
   d) a model train set

34. An antenna with 3dB gain is fed by a 25W rf signal. The electric field strength at 100m is approximately:
   a) 0.05V/m
   b) 0.1V/m
   c) 0.5V/m
   d) 3.5V/m

35. The fifth harmonic of a 145MHz transmission lies in:
   a) an amateur band
   b) a vhf band
   c) a microwave band
   d) a uhf tv band

36. To minimise rf going back into the mains, which of the following could be fitted in the mains input of a piece of equipment?

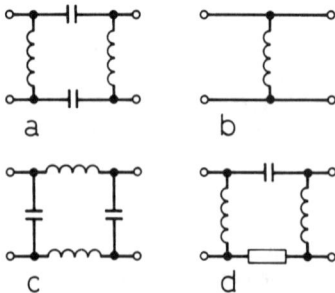

37. A neighbour complains of interference to their television but says it goes when they disconnect the antenna. This coincides with your transmission times. As a first step:
   a) try a mains filter
   b) suggest they use a set top antenna
   c) try a filter in the tv down lead
   d) renew the antenna cable

38. In which of the following will the RIS offer a visit or diagnosis:
   a) interference on a rental television
   b) interference on a car radio
   c) interference on an owners tv
   d) interference on an owners set using a tv set-top antenna

39. The main cause of intermodulation products in a receiver is produced by:
   a) the receiver being tuned off channel
   b) a crystal filter being used
   c) non-linearity in the rf stages
   d) a good preselector being used

40. In fitting a filter in a tv downlead which of the following combinations are correct?

	Impedance	Insertion Loss
a)	75 ohms	3dB
b)	3 ohms	75dB
c)	25 ohms	60dB
d)	50 ohms	–75dB

41. Which of the following filters should be used in order to attenuate 21MHz signals on a tv down lead?
   a) a low pass filter
   b) a high pass filter
   c) a band reject filter at the tv frequency
   d) a band pass filter at 21MHz

42. It is found that high erp on 144MHz causes tv interference in a block of flats. Apart from transmitter output which of the following would keep rf field strength to a minimum?
   a) an 11 element beam
   b) stacked 4 over 4, 11 element beams
   c) a 5 element beam
   d) an 8 over 8 skeleton slot fed beam

43. Interference is only caused on a hi-fi system when the audio cassette deck is in use. This might be eliminated by:
   a) a ceramic capacitor across a base emitter junction in the cassette player preamplifier
   b) an electrolytic capacitor across a base emitter junction in the cassette player preamplifier
   c) additional power supply smoothing
   d) a 1 Mohm resistor across a base emitter junction in the cassette player preamplifier

44. Severe interference on a radio receiver is experienced when a pcb with digital circuits is running on the bench. This interference might be reduced if:
   a) the pcb was put in a plastic box

b) the pcb was encased in epoxy resin
c) the pcb was placed in an earthed wooden box
d) the pcb was placed in a screened copper box

45. Next door's baby alarm is connected by a long length of twin flex. It is prone to interference from a nearby transmitter. To minimise unwanted pick up one could suggest:
a) using double the length of flex
b) the use of screened connecting cable
c) the use of a double twin flex
d) splitting the twin flex and running via different routes

Sample examination 4, Paper 2
Operating practices, procedures and theory.

1. In a contest always:
a) give a 59 report
b) give an accurate report
c) refuse to give a report
d) reply in cw only

2. Using morse telegraphy, WX is used to denote:
a) how do you copy?
b) weather
c) how many watts?
d) will you repeat

3. The purpose of a terrestrial repeater is to:
a) increase satellite coverage
b) increase the range of mobile stations
c) increase the range of fixed stations
d) minimise contacts of pedestrian stations

4. Using the International Phonetic Alphabet, RADIO would be:
a) romeo, alpha, delta, india, oscar
b) radio, alpha, delta, india, oscar
c) romeo, alpha, denmark, india, oscar
d) radio, alpha, delta, italy, oscar

5. When calling a station, it is good practice to:
a) put your callsign first
b) use your callsign only
c) put the callsign of the station being called first
d) use the callsign of the other station only

6. In the RST code, T stands for:
a) temperature
b) tone
c) time
d) transmitter

7. Using the Q code, interference from other stations is referred to as:
a) QRI
b) QRM
c) QRN
d) QRO

8. If a transmission receives a readability report of R5, it is:
a) unreadable
b) readable with considerable difficulty
c) readable with practically no difficulty
d) perfectly readable

9. When using cw, the invitation for a specific station to transmit is:
a) BT
b) K
c) KN
d) T

10. What is the total capacitance in the above circuit?
a) 0.15nF
b) 2.7nF
c) 4.5nF
d) 202.5nF

11. Two resistors are connected in series across a dc supply. The voltage across one resistor is 12V and the other 24V. The supply voltage is:
a) 6V
b) 12V
c) 24V
d) 36V

12. As frequency rises the reactance of a capacitor:
a) stays constant
b) decreases
c) increases
d) goes infinite

13. A tuned circuit can be made to oscillate providing:
a) additional damping is included
b) more capacitance is added
c) more inductance is added
d) an amplifier is added in order to make up for circuit losses

14. The total resistance in the above circuit is:
a) 250 ohms
b) 500 ohms
c) 1 kohm
d) 4 kohms

15. The quiescent current of a Class C amplifier is:
a) zero
b) infinite
c) the same as the full load value
d) just greater than zero

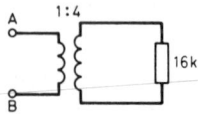

16. The reflected impedance seen at terminals AB in the above diagram is:
    a) 0 ohms
    b) 250 ohms
    c) 1000 ohms
    d) 4000 ohms

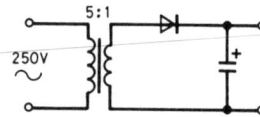

17. The above circuit is typical of:
    a) an L-C oscillator
    b) a common base stage
    c) a Class C tuned amplifier
    d) a crystal oscillator

18. Integrated circuits that perform logic functions come under the general classification of:
    a) linear circuits
    b) amplifiers
    c) mixers
    d) digital circuits

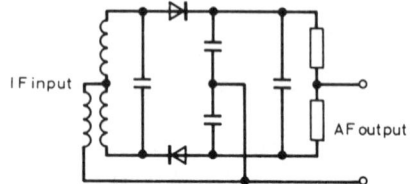

19. The output voltage between A and B will be about:
    a) 2.2V
    b) 5V
    c) 10V
    d) 13V

20. Referring to the circuit of Q19, the purpose of the 10μF capacitor is:
    a) to smooth the voltage across the Zener
    b) to remove high voltage spikes
    c) prevent transistor overloading
    d) for rf decoupling

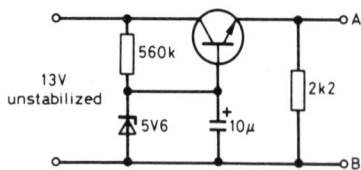

21. The piv rating of the diode in the above circuit should not be less than:
    a) 50V
    b) 75V
    c) 150V
    d) 250V

22. The phase relationship between input and output voltage of a common emitter stage is:
    a) zero
    b) 90 degrees
    c) 180 degrees
    d) 270 degrees

23. A receiver that converts all incoming signals to a fixed i.f is known as:
    a) a superheterodyne receiver
    b) a synchrodyne receiver
    c) a direct conversion receiver
    d) a crystal receiver

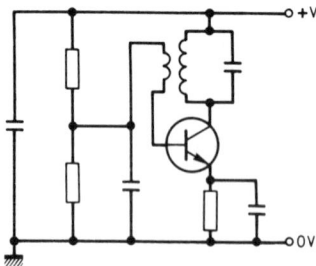

24. The above circuit is:
    a) an envelope detector
    b) a crystal detector
    c) a ratio detector
    d) a varactor multiplier

25. The circuit in Q24 is used for:
    a) detecting a.m signals
    b) demodulating fm signals
    c) product detecting ssb signals
    d) detecting vestigial sideband signals

26. A low power transmitter feeds an amplifier that has a gain of 10dB. If the input power to the amplifier is 1W, the output is:
    a) 1dBW
    b) 3dBW
    c) 10dBW
    d) 20dBW

27. A Class C amplifier delivers 100W of cw to the antenna. The input dc power is about:

a) 100W
b) 120W
c) 150W
d) 400W

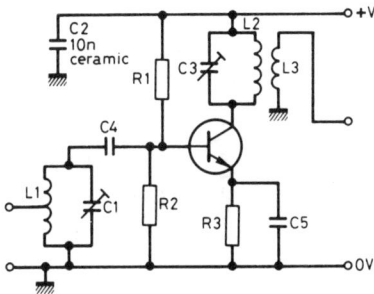

28. The above circuit is that of a:
a) mixer
b) rf amplifier
c) af amplifier
d) oscillator

29. The purpose of C2 in the circuit of Q28 is:
a) af decoupling
b) tuning
c) ac coupling
d) rf decoupling

30. The tapping on L1 in the circuit of Q28 is:
a) for impedance matching
b) ac coupling only
c) dc coupling only
d) local oscillator input

31. To minimise degradation of the selectivity of the circuit in Q28, the turns of L3 should be:
a) much greater than those on L2
b) equal to those on L2
c) much less than those on L2
d) one turn greater than those on L2

32. The typical efficiency of an hf linear operating in Class AB is:
a) 35%
b) 45%
c) 55%
d) 65%

33. An oscillator should ideally be followed by:
a) a buffer
b) a power amplifier
c) a Class C amplifier
d) a notch filter

34. Frequency modulation has theoretically:
a) an infinite number of sidebands
b) only two sidebands
c) no sidebands
d) only one sideband

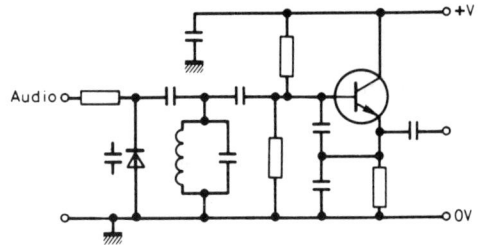

35. The above circuit is:
a) an amplitude modulator
b) an ssb generator
c) a demodulator
d) a frequency modulator

36. If a transmitter output impedance is 50 ohms, for optimum power transfer the load should be:
a) 50 ohms
b) 75 ohms
c) 100 ohms
d) 150 ohms

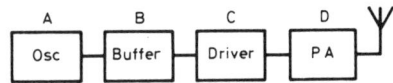

37. The above is a cw transmitter. Keying should be applied at:
a) A
b) B
c) C
d) D

38. The angle between the E and H fields in an electromagnetic wave is:
a) 45 deg
b) 90 deg
c) 180 deg
d) 270 deg

39. Which of the following cables has the lowest rf losses?
a) twisted flex
b) coaxial cable
c) open wire feeder
d) mains cable

40. The above antenna represents:
a) an end fed wire
b) a simple dipole
c) a cubical quad
d) a trap dipole

41. The radiation resistance of a half wave dipole is

about:
a) 50 ohms
b) 73 ohms
c) 100 ohms
d) 300 ohms

42. The velocity factor of open wire feeder is about:
a) 0.1
b) 0.4
c) 0.6
d) unity

43. The height of the F1 layer is about:
a) 80km
b) 120km
c) 200km
d) 300km

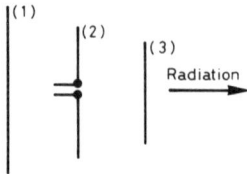

44. The above represents a simple beam. The names of the elements in numerical order are:
a) radiator, director, reflector
b) director, radiator, reflector
c) director, reflector, radiator
d) reflector, radiator, director

45. A balun is:
a) a balance to unbalance transformer
b) a mains transformer
c) a single winding inductor
d) a semiconductor device

46. The critical frequency is:
a) the highest frequency reflected when the radiation is vertical
b) the lowest frequency reflected when the radiation is vertical
c) the highest frequency reflected when the radiation is horizontal
d) the lowest frequency reflected when the radiation is horizontal

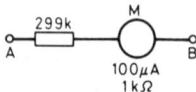

47. To obtain full scale deflection on the meter, the dc voltage across AB must be:
a) 1V
b) 5V
c) 10V
d) 30V

48. A 50 ohm dummy load is made from eleven 560 ohm

carbon resistors each of 5W rating. Total safe power that can be dissipated is:
a) 0.5W
b) 5W
c) 27.5W
d) 55W

49. Which of the circuits below could be used to estimate the value of a resistor by Ohm's Law ?

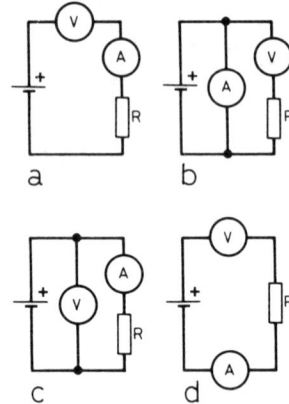

50. An swr meter is inserted into a perfectly matched transmitter/antenna system. The value shown should indicate:
a) 10W reflected power
b) 1:1 vswr
c) 1:0 vswr
d) 0:1 vswr

51. For a moving coil meter to respond to ac it must be combined with:
a) a dc blocking capacitor
b) a transformer
c) a bi-directional switch
d) a rectifier

52. An oscilloscope shows a peak to peak reading across a 1000 ohm resistor of 25V. The rms current through the resistor is:
a) 8.8mA
b) 12.5mA
c) 25mA
d) 50mA

53. When first testing a transmitter it should be fed into:
a) an antenna
b) a capacitor of reactance 50 ohms
c) a 50 ohm wirewound resistor
d) a non-reactive 50 ohm dummy load

54. A frequency counter has a quoted accuracy of one part in a million. If it is on a range of maximum reading 100MHz, then the accuracy at the top of the range is to the nearest:
a) 1Hz

b) 10Hz
c) 100Hz
d) 1000Hz

55. The rms calibration of a rectifier instrument is:

a) only true for a sinewave
b) true for all waveforms
c) true for a square wave and a sinewave only
d) only true for harmonics

Sample examination 5, Paper 1
Licensing conditions, transmitter interference and electromagnetic compatibility

1. The callsign issued to a station resident in the Isle of Wight is:
   a) GW7xxx
   b) G7xxx
   c) GIW7xxx
   d) GI7xxx

2. The station log may be maintained:
   a) on a computer printout
   b) in a loose leaf binder
   c) on a magnetic disc
   d) in pencil

3. Which of the following is not eligible to operate an amateur radio station under the supervision of a class A licence holder?
   a) a class B licence holder
   b) an unlicensed member of the family
   c) another class A licence holder
   d) the holder of a Radio Amateurs' Examination Certificate

4. No part of a message in an amateur transmission must be:
   a) in plain language
   b) in a language other than the mother tongue
   c) sent in morse
   d) sent in secret code or cypher

5. The maximum carrier power supplied to the antenna in the 144MHz band is:
   a) 10dBW
   b) 15dBW
   c) 20dBW
   d) 26dBW

6. The transmission frequency must be determined by:
   a) a crystal only
   b) a frequency synthesiser only
   c) a phase locked loop only
   d) any satisfactory method

7. The 10m band lies between:
   a) 28.0 - 28.7MHz
   b) 28.7 - 29.0MHz
   c) 28.0 - 29.0MHz
   d) 28.0 - 29.7MHz

8. When using phonetics in telephony these can be:
   a) of an obscene nature
   b) formed from well known acceptable words
   c) of a misleading nature
   d) swear words

9. Tests for harmonics and spurious emissions shall be:
   a) tape recorded
   b) put in a loose leaf book
   c) recorded in the log
   d) undertaken daily

10. The age limit for a person with a City and Guilds Radio Examination Certificate operating under supervision is:
    a) not specified
    b) over 12 years of age
    c) over 14 years of age
    d) over 16 years of age

11. You are operating your station from a private sailing boat on the Norfolk Broads. Which suffix should be added to your callsign?
    a) M
    b) P
    c) MM
    d) A

12. Pulse emissions must:
    a) not be used
    b) be on frequency bands above 1000MHz
    c) be on frequency bands below 1000MHz
    d) be used with a peak power which exceeds the p.e.p

13. Single sideband, suppressed carrier transmission is denoted by:
    a) A3E
    b) F3E
    c) R3E
    d) J3E

14. If an urgent request for medical drugs is received from anywhere, this should be passed to:
    a) the local St John's Ambulance Brigade
    b) the local Red Cross branch
    c) the local doctor
    d) a duly authorised officer of Her Majesty's Government

15. Which of the following bands is allocated to amateurs on a secondary basis?
    a) 1.81 - 2.0MHz
    b) 3.5 - 3.8MHz
    c) 10.1 — 10.15MHz
    d) 144 - 146MHz

16. Tests should be made on one's equipment to check for harmonic radiation. These should be:
    a) from time to time
    b) every 3 months
    c) every time the licence is renewed
    d) weekly

17. To prevent unwanted radiation in the shack rf connections between units should be by:
    a) open wire feeder
    b) good quality coaxial cable
    c) bell wire
    d) mains type cable

18. A frequency synthesiser in a transmitter consists of a vco, digital dividers, phase locked loop and so forth, it should be:
    a) encased in a warm plastic enclosure
    b) placed in the p.a compartment
    c) enclosed in a screened box
    d) fed by open wire feeder

19. When using a digital frequency counter to check the calibration of a transmitter output signal:
    a) connect the counter to the vfo or synthesiser output
    b) use only a carrier with no modulation
    c) key the transmitter at 12 wpm
    d) use a frequency modulated carrier

20. Which of the following circuits would be useful in the output of an hf transmitter in order to minimise harmonic radiation?

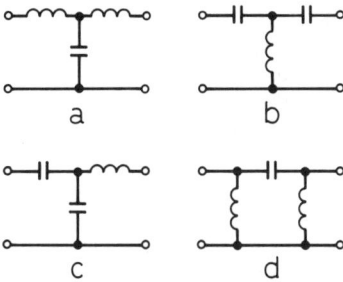

21. Which of the following might be effective at reducing the risk of parasitic oscillations in a low power vhf output stage:
    a) ferrite beads on the emitter lead of the power device
    b) ferrite beads on the microphone cable
    c) ferrite bead in series with microphone
    d) ferrite bead on loudspeaker lead

22. To minimise unwanted radiation of sub-harmonics and harmonics, a vhf transmitter should be followed by:
    a) a low pass filter
    b) a band pass filter
    c) a high pass filter
    d) a notch filter

23. The maintenance of a transmitter shall be such that:
    a) it causes no undue interference
    b) it is on the verge of self oscillation
    c) parasitic oscillations are present
    d) all internal shielding is removed

24. If a transmitter is over modulated or overdriven, it is likely to cause:
    a) harmonics
    b) sub-harmonics
    c) a change in modulation mode
    d) small dc variations

25. To check for harmonics in a radiated signal, which of the following could be used?
    a) an swr meter
    b) an absorption wavemeter
    c) a digital frequency meter
    d) a dip meter

26. To keep the bandwidth of an fm transmission to about that for a.m, the modulation index should be kept to about:
    a) 0
    b) 0.6
    c) 5
    d) 15

27. When a transmitter is not keyed but has dc power applied to it, the ammeter on the power supply shows a small but varying current. This might indicate:
    a) the presence of parasitic oscillations
    b) it is ready for the morse key to be plugged in
    c) there is interference on the mains
    d) the presence of parakeet noise

28. If a Class AB linear amplifier is overdriven by an ssb signal it will:
    a) cause splatter on adjacent frequencies
    b) go into parasitic oscillation
    c) introduce a carrier
    d) produce a clean signal

29. The following should always be used with a morse key:
    a) a transistor amplifier
    b) no filtering
    c) a notch filter
    d) a key-click filter

30. When checking the output of an fm transmitter the third harmonic with respect to the carrier should be at least:
    a) −6dB
    b) −16dB
    c) −30dB
    d) −100dB

31. When trying to rf decouple a loudspeaker lead the type of capacitor to be used is:
    a) ceramic
    b) aluminium electrolytic
    c) tantalum
    d) polycarbonate

32. A domestic receiver having an i.f of 455kHz and receiving a signal on 945kHz experiences strong interference from someone on the 160m band. This could be caused by second channel interference on:
    a) 1.810MHz
    b) 1.825MHz
    c) 1.835MHz
    d) 1.855MHz

33. When transmitting in the 1.81 - 2.0MHz band, second channel interference problems may occur on:
    a) the uhf tv band
    b) the fm broadcast band
    c) the long waveband
    d) the medium waveband

34. To keep the electric field below 1V/m at a distance of 100m, the maximum erp should be:
    a) 14W
    b) 100W
    c) 200W
    d) 400W

35. It is found that interfering signals are being induced on the braid of an antenna down lead to a domestic fm broadcast receiver by a 144MHz transmitter. One possible method of reducing the interference is:
    a) to fit a "braid breaker"
    b) remove the 144MHz earth connection
    c) increase the 144MHz output power
    d) fit the 144MHz transmitter with a low pass filter

36. The use of an indoor transmitting antenna may be considered on aesthetic grounds but:
    a) it will give only long distance contacts
    b) it has a greater chance of coupling into the mains wiring
    c) the planners will object
    d) will attract lightning

37. A 435MHz transmitter/antenna system gives 1kW erp and points straight at a neighbour's tv antenna. This could cause:
    a) problems with the 435MHz receiver
    b) problems with the 435MHz transmitter
    c) self oscillation of the transmitter
    d) overloading of the tv front end

38. To remove a narrow band of frequencies that may cause interference, it might be possible to use:
    a) a notch filter
    b) a mesh filter
    c) a resistor
    d) an all stop filter

39. A neighbour experiences interference on a tv from a nearby radio transmitter. Bell flex has been used for the tv down lead. As a first attempt:
    a) replace the flex with a good quality coaxial cable
    b) run two sets of bell flex in parallel
    c) use mains cable instead of bell flex
    d) use a tv preamplifier

40. Which of the following devices is prone to causing wideband noise ?
    a) a home computer
    b) a 1kHz tone generator
    c) an rf generator set to 90MHz
    d) a 28MHz transmitter

41. If a ferrite bead is to be fitted on a transistor lead to reduce an interference problem, which of the following is most likely to produce a cure?

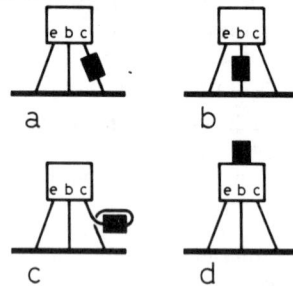

42. RF interference is found to be entering a tv by the i.f lead. It is impractical to remove the lead from the tuner, hence one solution is to put a capacitor between screen and centre of the lead. The most likely choice is:
    a) a 10μF electrolytic
    b) a 10nF disc ceramic
    c) a 2pF polystyrene
    d) a 1μF polycarbonate

43. The coaxial cable from a transmitter feeds a balanced half wave dipole. The cable runs vertically next to the house and causes some interference. To minimise this problem it would be wise to:
    a) replace coaxial cable with a single wire feeder
    b) put loops into the coaxial cable
    c) put the coaxial cable in a plastic pipe
    d) feed the dipole via a balun

44. A transmitter has a fixed power output and fixed length of cable to the antenna. To keep the erp reasonably low without adversely affecting receive one should:
    a) put resistors in series with the coaxial cable
    b) put chokes in series with the coaxial cable
    c) use an antenna with low gain
    d) use plastic antenna elements

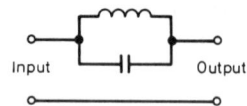

45. The filter shown above is:
    a) a band stop filter
    b) a low pass filter
    c) a high pass filter
    d) a resistive pad

Sample examination 5, Paper 2
Operating practices, procedures and theory.

1. Using the International Phonetic Alphabet DIODE would be:

a) delta, india, oscar, delta, echo
b) denver, india, oscar, denver, echo
c) delta, italy, oscar, delta, echo
d) diode, india, oscar, diode, echo

2. To be able to operate through a satellite, one must:
a) use an access tone
b) get permission from AMSAT
c) call the satellite with its number
d) just transmit in the correct frequency range

3. Before making a CQ call:
a) listen on the frequency before commencing
b) send a series of Vs
c) send a 1750Hz tone
d) keep giving your callsign

4. The Q code for 'Who is calling me?' is:
a) QRW
b) QRZ
c) QRQ
d) QRX

5. For safety the integrity of the earthing system should be checked:
a) every decade
b) when you move house
c) periodically
d) never

6. When testing on live equipment it is good practice to:
a) keep one hand in a pocket
b) use uninsulated probes
c) have both hands in the equipment
d) use a soldering iron

7. The main purpose of a repeater is:
a) to improve communication between mobile stations
b) to provide a convenient frequency for local nets
c) to aid dx working during lift conditions
d) to aid propagation studies

8. When using telephony, it is good practice to:
a) use the phonetic alphabet as much as possible
b) always use Q codes
c) speak clearly and not too quickly
d) make long transmissions before allowing the other station to reply

9. The letter A after an RST report indicates:
a) aurora
b) key clicks
c) chirp
d) drift

10. The power dissipated in a pure inductor is:
a) zero
b) small
c) large
d) very large

11. A light bulb is rated at 12V, 3W. The current drawn when illuminated is:
a) 250mA
b) 750mA
c) 4A
d) 36A

12. 100mW is equivalent to:
a) 0.001W
b) 0.01W
c) 0.1W
d) 1.0W

13. Two 10 kohm resistors are connected in parallel across a 5V dc supply. Total current taken is:
a) $50\mu A$
b) 0.5mA
c) 1mA
d) 1A

14. Two $10\mu F$ capacitors are placed in parallel across a 10V, 1kHz supply. The phase difference between applied voltage and current drawn is:
a) 0 deg
b) 45 deg
c) 60 deg
d) 90 deg

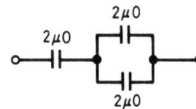

15. The total capacitance in the above circuit is:
a) $1.33\mu F$
b) $3\mu F$
c) $3.5\mu F$
d) $6\mu F$

16. A Zener diode has a sharp 'knee' in its:
a) forward bias characteristic
b) construction
c) oscillatory mode
d) reverse bias characteristic

17. In the diagram below, which represents the diode in a conducting condition ?

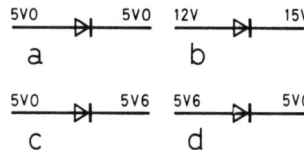

18. P type material has:
a) an excess of electrons
b) a deficiency of an atom
c) additional electrons
d) a deficiency of electrons

19. Colpitts, Hartley, Vackar, Clapp-Gouriet are all types of:
    a) power supply
    b) amplifier
    c) oscillator
    d) modulator

20. A varactor diode acts like:
    a) a variable resistance
    b) a variable voltage regulator
    c) a variable capacitance
    d) a variable inductance

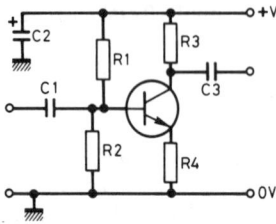

21. The above circuit is:
    a) an emitter follower
    b) a common emitter amplifier
    c) a common collector amplifier
    d) a voltage stabiliser

22. In the circuit of Q21, the purpose of R1 and R2 is:
    a) biasing
    b) amplification control
    c) feedback
    d) filtering

23. The frequency stability of a receiver is its ability to:
    a) stay tuned to the desired signal
    b) track the incoming signal as it drifts
    c) provide a frequency standard
    d) provide a digital readout

24. The typical bandwidth of a good filter for cw is about:
    a) 100Hz
    b) 500Hz
    c) 3kHz
    d) 10kHz

25. In which of the following type of single mode Rx is an agc system not necessary?
    a) a.m
    b) ssb
    c) cw
    d) fm

26. The tuned radio frequency receiver:
    a) has a severe second channel problem
    b) has three mixer stages
    c) does not have second channel problems
    d) is completely passive

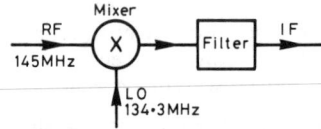

27. In the above arrangement, for an i.f less than the rf frequency, the filter should be tuned to:
    a) the average of the lo and rf frequencies
    b) the sum of the lo and rf frequencies
    c) the difference in the rf and lo frequencies
    d) none of these

28. The type of filter used in Q27 would normally be referred to as:
    a) a bandpass filter
    b) a high pass filter
    c) a band stop filter
    d) a notch filter

29. To check the accuracy of the tuning dial a calibration oscillator is provided. This is:
    a) crystal controlled
    b) inductance controlled
    c) capacitance controlled
    d) resistance controlled

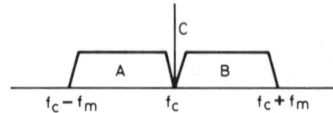

30. The above spectrum plot is typical of:
    a) amplitude modulation
    b) amplitude modulation, suppressed carrier
    c) single sideband, reduced carrier
    d) single sideband, full carrier

31. Referring to the diagram of Q30, what should be removed so as to represent a transmission using upper sideband only?
    a) B and C
    b) A
    c) A and C
    d) A and B

32. The rf spectrum required to transmit an ssb signal is:
    a) half the modulating signal bandwidth
    b) the same as the modulating signal bandwidth
    c) twice the modulating signal bandwidth
    d) always 3kHz

33. A filter is often fitted between a transmitter and the antenna at hf. It is usually:
    a) a low pass filter
    b) a high pass filter
    c) an i.f filter
    d) an audio filter

34. At 3.5MHz, a wire 40 metres long corresponds to:
    a) quarter wavelength

b) half wavelength
c) one wavelength
d) two wavelengths

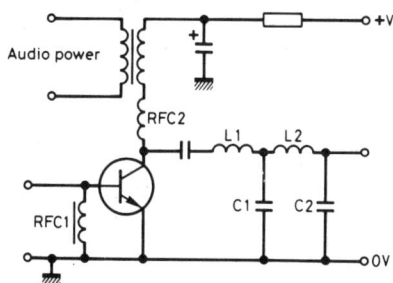

35. The above circuit is that of:
    a) a frequency modulator
    b) an amplitude modulator
    c) a phase modulator
    d) an amplitude demodulator

36. In the circuit of Q35, the components L1, L2, C1 and C2 provide:
    a) harmonic suppression and output impedance matching
    b) harmonic suppression only
    c) output matching only
    d) audio filtering

37. In the circuit of Q35, RFC2:
    a) provides resonance at audio frequencies
    b) prevents rf getting back into the modulator
    c) is a superfluous component
    d) is to damp the filter

38. A crystal oscillator can:
    a) be varied over a very wide range
    b) be always unstable
    c) be kept in a draught
    d) only be varied over a very small range

39. The major mode of propagation up to about 2MHz is:
    a) by direct wave
    b) by tropospheric wave
    c) by ionospheric wave
    d) by ground wave

40. A beam is used as shown above, the transmitted signal will be:
    a) vertically polarised
    b) horizontally polarised

c) elliptically polarised
d) non-polarised

41. In electromagnetic radiation, which of the following is true?
    a) E and H are at 180 degrees to each other
    b) E, H and the direction of propagation are all at right angles to each other
    c) the angle between E and H is zero
    d) the direction of propagation is at 180 degrees to the E field but in line with the H field

42. The addition of reflectors and directors to a folded dipole:
    a) raises its radiation resistance
    b) has no effect on its radiation resistance
    c) lowers its radiation resistance
    d) will not provide any directivity

43. The bandwidth of a beam antenna is dependent on:
    a) radiation resistance of the dipole
    b) spacing of directors and reflectors
    c) feed cable impedance
    d) propagation conditions

44. The velocity factor of a coaxial cable with solid polythene dielectric is about:
    a) 0.66
    b) 0.1
    c) 0.8
    d) 1.0

45. Refraction of an electromagnetic wave is:
    a) the same as reflection
    b) the bending of its path
    c) absorption by the ionosphere
    d) bouncing from a stellar object

46. A half wavelength transmission line is terminated with 68 ohms, the input to the line is:
    a) 34 ohms
    b) 68 ohms
    c) 136 ohms
    d) 204 ohms

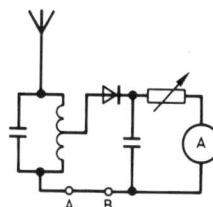

47. The above circuit is that of:
    a) an absorption wavemeter
    b) a dip oscillator
    c) an swr meter
    d) heterodyne wavemeter

48. If headphones are inserted between A and B in the circuit of Q47, it allows monitoring of:

a) a.m signals
b) fm signals
c) harmonics
d) parasitic oscillations

49. An swr meter should be placed between:
a) coaxial cable and antenna
b) atu and antenna feeder
c) transmitter and atu
d) none of these

50. Which of the following would be used to examine the shape of a waveform?
a) an oscilloscope
b) an absorption wavemeter
c) a digital frequency counter
d) a dip meter

51. To check that a crystal is working on its correct overtone, the simplest piece of equipment necessary is:
a) a voltmeter
b) an ammeter
c) an absorption wavemeter
d) a dip oscillator

52. For accuracy a digital frequency meter should be based on:
a) an R-C timebase
b) an L-C timebase
c) a crystal timebase
d) an R-L timebase

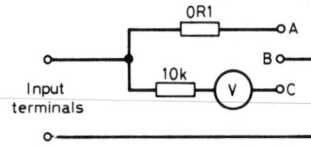

53. To use the above arrangement as a voltmeter, which connection should be made?
a) A to B
b) C to A
c) C to B
d) A to B to C

54. How much current will be taken by a 20 kohm/V meter for full scale deflection?
a) $50\mu A$
b) $100\mu A$
c) $200\mu A$
d) $500\mu A$

55. The coaxial cable from an swr meter to an antenna at hf develops a fault so that no power reaches the antenna. The swr meter will read:
a) zero
b) 1:1
c) high
d) very low

Sample examination 6, Paper 1
Licensing conditions, transmitter interference and electro-magnetic compatibility.

1. A class B licence does not authorise:
    a) use of frequencies below 30MHz
    b) use of frequencies above 144MHz
    c) pulse modulation
    d) microwave working

2. The required speed for the morse test is:
    a) 10 wpm
    b) 12 wpm
    c) 15 wpm
    d) 20 wpm

3. The callsign of an amateur resident in Guernsey with a class A licence might be:
    a) GU6xxx
    b) GU3xxx
    c) GC6xxx
    d) GJ2xxx

4. When maritime mobile, the licensee must cease to operate on the demand of:
    a) the coastguard
    b) the vessel's master
    c) the vessel's radio operator
    d) the harbour master

5. A licensee must be able to:
    a) use two different languages
    b) verify that his transmissions are within the authorised frequency band
    c) read morse at 20 wpm
    d) write

6. The maximum peak envelope power permitted for ssb operation on the 14MHz band is:
    a) 15dBW
    b) 20dBW
    c) 26dBW
    d) 30dBW

7. In the microwave bands, the maximum permitted power density is:
    a) $10mW/cm^2$
    b) $15mW/cm^2$
    c) $20mW/cm^2$
    d) not specified

8. Which of the following is issued to amateurs on a secondary basis?
    a) 7.0 - 7.1MHz
    b) 14.0 - 14.35MHz
    c) 144 - 146MHz
    d) 430 - 440MHz

9. A Northern Ireland station when mobile in Wales should sign as:
    a) GIxxx/GW/M
    b) GWxxx/GI/M
    c) GIxxx/M
    d) GWxxx/M

10. Which of the following recorded messages may be transmitted by an amateur station?
    a) instructions on the use of a repeater
    b) the RSGB news
    c) a continuous audio tone
    d) a public broadcast

11. In filling in the log book, one should not:
    a) enter time in UTC
    b) write in indelible pencil
    c) put the callsign of the other station
    d) leave gaps

12. The licence prohibits the use by amateurs of the allocation 430 - 432MHz in an area in:
    a) Devon and Cornwall
    b) the north of England
    c) Scotland
    d) Northern Ireland

13. When pedestrian which of the following suffixes should be used?
    a) /A
    b) /T
    c) /P
    d) /M

14. Which of the following is an infringement of the licence conditions?
    a) operation from a private launch on Lake Windermere
    b) operation from a private coach on the M90
    c) operation from a private plane
    d) operation from a bicycle

15. Fast scan tv is allowed on which of the following?
    a) all bands
    b) 144MHz and above
    c) 144MHz and below
    d) where indicated on 430MHz and above

16. Which of the following is a high pass filter?

17. A digital frequency counter shows 7.0885MHz when

checking a 7MHz transmission. The figure 5 represents:
a) units of Hertz
b) tens of Hertz
c) hundreds of Hertz
d) thousands of Hertz

18. The simplest piece of equipment to check the correct harmonic selection in a multiplier stage is:
a) a multimeter
b) a diode probe
c) an absorption wavemeter
d) a digital frequency counter

19. A transmitter is amplitude modulated using a carrier frequency of 21.448MHz. The speech bandwidth is 3kHz. The transmission will:
a) comply with the licence schedule
b) cause splatter
c) cause harmonics
d) exceed the band edge

20. To minimise the radiation of one particular harmonic one can use a:
a) trap in the transmitter output
b) resistor
c) high pass filter in the transmitter output
d) filter in the receiver lead

21. Power supply leads in a transmitter should:
a) be well decoupled at rf
b) not be filtered
c) be routed by the p.a compartment
d) have rf oscillations on them

22. Chirp is a form of frequency instability. It is caused by:
a) background noise
b) overmodulation
c) over deviation
d) pulling of an oscillator when keying

23. A spurious emission from a transmitter is:
a) an unwanted emission that may be harmonically related to the carrier
b) an unwanted emission that is harmonically related to the modulating audio frequency
c) generated at 50Hz
d) the main part of the modulated carrier

24. The station shall be so constructed and maintained so as:
a) not to cause any undue interference to any wireless telegraphy
b) to cause interference with wireless telegraphy
c) to cause interference with tvs only
d) to produce harmonic and spurious radiations

25. To obtain high frequency stability in a transmitter, the vfo should:
a) be run from a non-regulated ac supply

d) be in a plastic box
c) be powered from a stable dc supply
d) change frequency with temperature

26. Key clicks in a cw transmission are caused by:
a) vfo instability
b) a very stable vfo
c) sharp edges to the carrier waveform
d) too slow a keying speed

27. The frequency changing stages in a transceiver should ideally be:
a) allowed to radiate freely
b) frequency modulated
c) encased in a waxed paper box
d) well screened to minimise unwanted radiation

28. To check the calibration of a transceiver with a vfo at the band edges, the minimum equipment required is:
a) a dip meter
b) a crystal controlled digital frequency meter
c) an absorption wavemeter
d) an oscilloscope

29. To minimise interference to adjacent channels, voice frequencies should be kept below:
a) 500Hz
b) 1kHz
c) 3kHz
d) 5kHz

30. If interference is caused to a government wireless station there can be an oral demand to close down. This is followed initially by:
a) confirmation in writing
b) a visit from the police
c) confiscation of equipment
d) imprisonment

31. An amateur transmission on 1.85MHz causes second channel interference to a neighbour's radio receiving on 940kHz and having an i.f of 455kHz. The local oscillator of the receiver must be:
a) 585kHz
b) 910kHz
c) 1.395MHz
d) 2.305MHz

32. A 144MHz transmitter causes problems with a nearby fm broadcast band receiver. One likely problem is:
a) overloading of the receiver front end
b) harmonic radiation
c) over modulation
d) speech processing

33. A corroded connector on a receiving antenna connecting cable may cause:
a) unwanted mixing products due to it exhibiting non-linear characteristics
b) mains modulation

c) enhanced signal reception due to its frequency changing properties
d) increased pre-amplification

34. The harmonic from a 430 - 440MHz transmitter most likely to cause interference to the uhf band is:
a) the second
b) the third
c) the fourth
d) the fifth

35. Interference is being caused when you transmit at 144.35MHz on a neighbour's tv which uses a set top antenna. There is no problem on your tv which uses a 10 element antenna on the chimney. A possible cure is:
a) an external antenna for your neighbour's tv
b) a set top antenna for your tv
c) a preamplifier for your neighbour's tv
d) balanced feeder for your neighbour's set top antenna

36. The best place for an hf beam in order to minimise the possibility of interference for an amateur living in a semi-detached house is:
a) on the joint chimney stack in the centre of the pair of houses
b) overhanging next door's roof space
c) as high and as far away as possible
d) as low and as near the house as possible

37. To improve a station design with regards to emc one could:
a) bond the transmitter to the water pipe in the house
b) provide a good rf earth for all the equipment
c) remove all earth connections
d) connect to the incoming alkathene water pipe

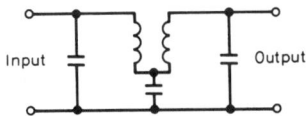

38. The above filter is an example of:
a) a low pass filter
b) a high pass filter
c) a band pass filter
d) a notch filter

39. A buzz occurs on tv sound along with white "tadpoles" across the screen. This interference is most likely to be caused by:
a) a taxi transmitter
b) a vacuum cleaner
c) a clockwork brush
d) an audio signal generator

40. When transmitting fast scan tv, your picture appears on a neighbour's tv. To remedy the situation one could:
a) put blue movies on
b) reduce power output to a minimum
c) increase modulation depth
d) increase picture contrast

41. A 25 watt fm transmitter feeds via very short cable an antenna with 3dB gain. At 50 metres this gives a field strength of about:
a) 0.1V/m
b) 0.5V/m
c) 1V/m
d) 7V/m

42. At the "notch" in a notch filter the insertion loss should be:
a) as high as possible
b) as low as possible
c) about 75 ohms
d) approximately 1.5MHz

43. Having fitted a high pass filter in a tv down lead little improvement is found with an interference problem. As a next practical step one could:
a) fit a braid breaker in the tv downlead
b) screen the inside of the tv with foil
c) earth the tv
d) remove the earth from the transmitter

44. It is well known that tv preamplifiers are:
a) very narrow band
b) good preselectors
c) very wide band
d) of zero gain

45. In helping a neighbour to set up a hi-fi system and locate the loudspeakers one should suggest:
a) long unscreened leads
b) long screened leads
c) screened leads as short as possible
d) the use of mains type cable

Sample examination 6, Paper 2

Operating practices, procedures and theory.

1. The 'golden rule' in morse telegraphy is:
a) never send faster than one can receive
b) send as fast as the transmitting station
c) send slower than one can receive
d) keep to 20 wpm

2. The Q code for closing down is:
a) QRT
b) QRC
c) QRZ
d) QRX

3. Before commencing transmission the operator should:
a) listen on the frequency to see if it is clear
b) turn the af gain down

c) turn the rf gain down
d) detune the antenna

4. Using the International Phonetic Alphabet, HENRY would be:
a) hotel, enrica, norway, romeo, yankee
b) hotel, echo, nancy, romeo, yokahoma
c) hotel, echo, november, romeo, yankee
d) hotel, echo, november, radio, yankee

5. The duration of an access tone for a UK repeater should be at least:
a) 100ms
b) 300ms
c) 1s
d) 5s

6. The use of repeaters by base stations:
a) should be encouraged
b) is illegal
c) will damage the repeater
d) should not be encouraged

7. Which of the following represents a fairly good signal strength using the RST code?
a) R5
b) S2
c) S5
d) S8

8. In order to operate through a satellite, it is necessary to:
a) transmit in the correct frequency range
b) use the maximum legal power
c) use a toneburst
d) obtain permission from the DTI

9. Band plans should:
a) be ignored
b) be observed because they are mandatory
c) be used only by new operators
d) be observed because they try to aid operating

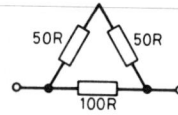

0·25ms/div

10. The above diagram represents a trace on an oscilloscope. What is the frequency of the displayed waveform?
a) 100Hz
b) 1kHz
c) 10kHz
d) 100kHz

11. Using the diagram of Q10, what is the peak to peak value of the waveform?
a) 1V

b) 2V
c) 10V
d) 20V

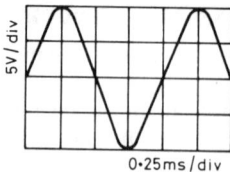

12. If there is 10V across the above resistors, what is the total current flow?
a) 0.002A
b) 20mA
c) 200mA
d) 2A

13. Kilo is equivalent to:
a) one thousandth
b) one thousand
c) one million
d) one thousand million

14. The Q factor of a 1mH coil with a winding resistance of 10 ohms at 100kHz is:
a) 6.28
b) 62.8
c) 628
d) 1000

15. The power dissipated by a 10 ohm resistor with 2A flowing through it is:
a) 5W
b) 20W
c) 40W
d) 200W

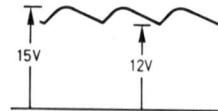

16. The above trace is that of the output of a power supply. The ripple is:
a) 1V
b) 3V
c) 12V
d) 15V

17. The fundamental output frequency from a full wave rectifier connected to a 50Hz supply with no smoothing is:
a) 25Hz
b) 50Hz
c) 100Hz
d) 200Hz

18. If the base voltage is increased on an npn transistor operating in Class B, the collector current will:
a) remain constant
b) decrease a little

c) increase noticeably
d) decrease noticeably

19. Which of the following circuits gives full wave rectification of the polarity shown?

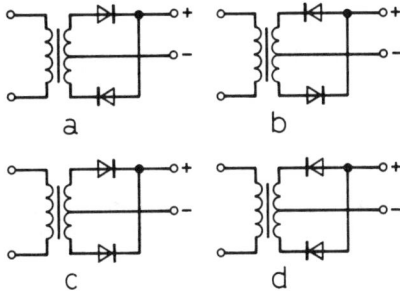

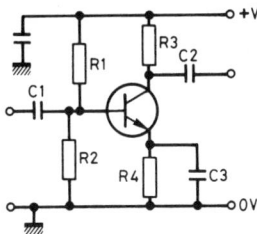

20. In the above circuit, C3 is for:
a) biasing
b) decreasing the amplifier gain
c) stopping dc drift
d) by-passing the emitter resistor

21. The purpose of C1 and C2 in the circuit of Q20 is:
a) dc coupling
b) ac coupling
c) biasing
d) tuning

22. If the capacitor C3 is removed from the circuit of Q20, the voltage gain will:
a) increase
b) stay the same
c) decrease
d) go to zero

23. Noise from ignition systems, thermostats, electrical machinery etc is more noticeable with:
a) fm signals
b) phase modulated signals
c) angle modulated signals
d) ssb signals

24. In a receiver, narrow bandwidth usually gives:
a) poor selectivity
b) no selectivity
c) high selectivity
d) negative selectivity

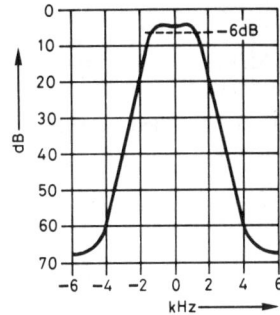

25. The above diagram represents the response of a filter. It would be most suitable in a receiver for:
a) wideband fm signals
b) a.m signals
c) fast scan tv signals
d) ssb signals

26. Referring to the diagram of Q25. the skirt bandwidth at –60dB is:
a) 3kHz
b) 5kHz
c) 8kHz
d) 9kHz

27. The pass band loss of the filter depicted in Q25 is:
a) 3dB
b) 6dB
c) 9dB
d) 12dB

28. A calibration oscillator in a receiver is based on:
a) a crystal
b) an L-C circuit
c) an R-C circuit
d) a transformer

29. If a preamplifier is fitted ahead of a receiver this can:
a) degrade the strong signal capabilities of the receiver
b) never cause any receiver overloading
c) provide better mains hum rejection
d) increase vfo stability

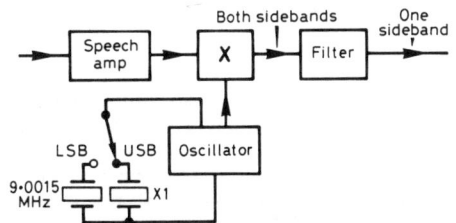

30. The above block diagram is that of an SSB generator. The filter is specified as 9MHz. What is the expected frequency of X1?

a) 8.9970MHz
b) 8.9985MHz
c) 9.0000MHz
d) 9.0030MHz

31. What is the box X called in Q30?
   a) a balanced modulator
   b) a balanced filter
   c) a balanced demodulator
   d) a non-balanced mixer

32. A typical bandwidth for the speech amplifier shown on the diagram in Q30, is:
   a) 6kHz
   b) 5kHz
   c) 4kHz
   d) 2.5kHz

33. The process of modulation allows:
   a) information to be impressed on to a carrier
   b) information to be removed from a carrier
   c) voice and cw to be combined
   d) none of these

34. The power output from an fm transmitter is:
   a) constant irrespective of modulation
   b) varies with the modulation
   c) is zero with no modulation
   d) reduces to 50% with modulation

35. For optimum stability a vfo should be:
   a) kept away from varying heat sources
   b) kept in a varying draught
   c) next to the power amplifier compartment
   d) air blown

36. Which of the traces below is that of a frequency modulated waveform?

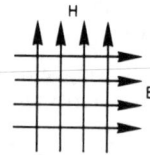

37. Which of the following spectrum diagrams is representative of the output of a balanced modulator?

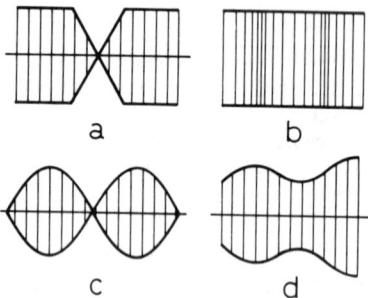

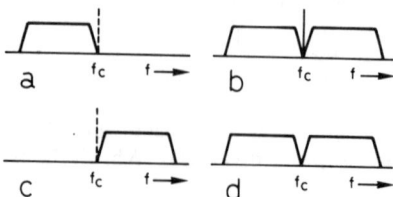

38. In the above diagram the polarisation is:
   a) vertical
   b) horizontal
   c) forwards
   d) backwards

39. To match a 300 ohm antenna to a 75 ohm transmission line, a transformer can be used that has a ratio of:
   a) 2:1
   b) 4:1
   c) 8:1
   d) 16:1

40. The unit of Z_0 is:
   a) ohms
   b) farads
   c) siemens
   d) henries

41. A half wave dipole will also resonate at:
   a) a sub-harmonic
   b) the second harmonic
   c) the third harmonic
   d) the fourth harmonic

42. Fading can be caused by:
   a) a poor antenna
   b) horizontal polarisation
   c) interaction of the sky and ground wave
   d) poor coaxial cable

43. Inserting traps into each leg of a dipole:
   a) allows it to operate only on one band
   b) cuts out harmonics
   c) gives broad band matching
   d) allows it to resonate on at least two bands

44. When the F2 layer is more highly ionised than usual, it will cause:
   a) poorer reflections at higher frequencies
   b) a lower m.u.f
   c) a higher m.u.f
   d) poor hf conditions

45. An antenna wire 15m long is to be used on the 3.5MHz band. What should be connected in series with it to make it resonate?
   a) a capacitor
   b) an inductor
   c) a capacitor and inductor in series
   d) a capacitor and inductor in parallel

46. The highest frequency that can be used between two stations on hf for satisfactory communication is

called the:
a) optimum working frequency
b) critical frequency
c) maximum usable frequency
d) penetration limiting frequency

47. A digital frequency meter can be used to:
a) measure harmonic content accurately
b) measure sideband content
c) measure frequency deviation
d) measure frequency accurately

48. An rtty signal requires a bandwidth of ±3kHz. A frequency counter of accuracy 1 part per million is used to check the frequency readout of the 145MHz transmitter. How close can the signal be transmitted to the upper band edge in order to ensure that the transmission is within the licence conditions?
a) 1.55kHz
b) 2.855kHz
c) 3.145kHz
d) 4.45kHz

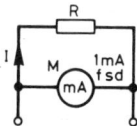

49. Using the above circuit to give an FSD for 1A, the current I must be:
a) 0.9A
b) 0.99A
c) 0.999A
d) 1.1A

50. If the resistance of the meter in Q49 is 500 ohms, the value of the shunt must be:
a) 0.05 ohm
b) 0.5 ohm
c) 5 ohm
d) 50 ohm

54. If there is ingress of water into a correctly matched antenna feeder, the vswr will:
a) tend to reduce
b) go below 1:1

c) stay constant
d) tend to rise

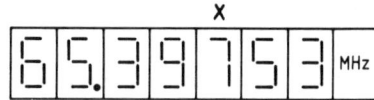

51. The above represents the output of a digital frequency counter. What does the digit at X give?
a) Hertz
b) tens of Hertz
c) hundreds of Hertz
d) thousands of Hertz

52. An ammeter in series with a 10 kohm resistor shows 5mA. The power dissipated by the resistor is:
a) 0.025W
b) 0.25W
c) 2.5W
d) 25W

53. To prevent altering the resonant frequency of a tuned circuit being examined with a dip oscillator, the coupling between the two should be:
a) very tight
b) resistive
c) fairly loose
d) the maximum possible

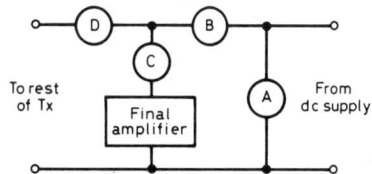

55. To measure dc input power to a final amplifier, meters are used. These should be:
a) voltmeter at A, ammeter at B
b) voltmeter at A, ammeter at C
c) voltmeter at C, ammeter at A
d) voltmeter at C, ammeter at D

Sample examination 7, Paper 1
Licensing conditions, transmitter interference and electro-magnetic compatibility.

1. The log must be retained for inspection for at least:
   a) 2 months after the last entry
   b) 1 year after the last entry
   c) 5 years after the last entry
   d) 6 months after the last entry

2. Which of the following bands are shared with other services?
   a) 3.5 - 3.8MHz
   b) 7.0 - 7.1MHz
   c) 14.0 - 14.35MHz
   d) 21.0 - 21.45MHz

3. When identifying using voice, the callsign must be in:
   a) plain language
   b) English
   c) facetious language
   d) morse code

4. When operating a handheld from a location on a hill top, one should use the suffix:
   a) /A
   b) /M
   c) /P
   d) /H

5. Telegraphy by on - off keying of a carrier without the use of a modulating audio frequency is designated:
   a) A1A
   b) A2A
   c) A2B
   d) A3E

6. The maximum power supplied to an antenna for cw at 144MHz must not exceed:
   a) 10dBW
   b) 15dBW
   c) 20dBW
   d) 26dBW

7. One method of notifying changes in the terms of the licence is by a notice in:
   a) the Edinburgh, Belfast and Cardiff Gazettes
   b) a newspaper published in each of London, Manchester and Glasgow
   c) a newspaper published in each of London, Douglas and Belfast
   d) the London, Edinburgh and Belfast Gazettes

8. Before operating an amateur station in a motor vehicle, it is necessary to:
   a) give the licensing authority the vehicle number
   b) inform the Vehicle Licensing Authority
   c) hold a current amateur licence
   d) obtain an additional licence from a Post Office

9. In order to operate maritime mobile an amateur must:
   a) obtain written permission from the vessel's master
   b) obtain a maritime mobile licence
   c) be the owner of the vessel
   d) operate only on cw

10. When in communication with another station the callsign must be sent:
    a) every 5 minutes
    b) every 10 minutes
    c) at least every 15 minutes
    d) at least every 30 minutes

11. In order to obtain a class A licence one must:
    a) have previously held a class B licence
    b) have passed the RAE and morse test
    c) have had at least one year's experience at sending morse
    d) have held a class B licence for one year and have passed the morse test

12. The callsign GW4xxx is issued to:
    a) a class A licensee living in Wales
    b) a class B licensee living in Wales
    c) a class A licensee living in Winchester
    d) a class A licensee living in Scotland

13. A station uses a transmitter that does not use crystals to determine the output frequency. The minimum frequency checking equipment should be:
    a) an absorption wavemeter
    b) based on a crystal reference
    c) an oscilloscope
    d) a multimeter

14. The term J3E in the schedule refers to:
    a) telephony, single sideband, reduced carrier
    b) telephony, single sideband, suppressed carrier
    c) telephony, double sideband
    d) telephony, single sideband, full carrier

15. If an antenna passes over an overhead power line, it must be guarded to the reasonable satisfaction of:
    a) the police
    b) the local council
    c) the RSGB
    d) the owner of the power line

16. When making a transmitter vfo, the coil should be:
    a) air cored
    b) placed in a position as free as possible from temperature variations
    c) placed next to a fan which cools the p.a
    d) wound as tightly as possible on a stainless steel former

17. Key clicks can be reduced by:
    a) slowing down the leading and falling edges of the waveform

b) slowing down the leading edge of the rf waveform only

c) slowing down the trailing edge of the rf waveform only

d) amplification

18. The total bandwidth of an fm transmission is found to be 15kHz. How close can the carrier be taken to the band edge in order not to produce out of band radiation?
   a) 0kHz
   b) 3kHz
   c) 7.5kHz
   d) 15kHz

19. If the bandwidth of an fm transmission is too wide it can be reduced by:
   a) the volume control
   b) the deviation control
   c) the bias control
   d) reducing the carrier frequency

20. A vhf/uhf transmitter should ideally be followed by:
   a) a high pass filter
   b) a crystal filter
   c) a band pass filter
   d) a mains filter

21. The presence of harmonics can be checked easily with:
   a) an absorption wavemeter
   b) an oscilloscope
   c) a digital frequency counter
   d) a diode probe

22. So as not to cause unnecessary sideband splatter the percentage modulation of an a.m signal should be kept below:
   a) 25%
   b) 50%
   c) 75%
   d) 100%

23. Which type of mixer keeps unwanted outputs to a minimum:
   a) balanced mixers
   b) product detectors
   c) single transistor mixers
   d) single diode mixers

24. Parasitic oscillations can cause interference. They are:
   a) always the same frequency as the mains supply
   b) always twice the operating frequency
   c) not related to the operating frequency
   d) three times the operating frequency

25. To stop unwanted radiations from an oscillator, it should be:
   a) enclosed in a metal box
   b) left unscreened

c) not be rf decoupled

d) placed in a paper box

26. To minimise the risk of unwanted radiations even with a matched antenna system, it is wise to:
   a) have a high swr
   b) use an atu or filter
   c) monitor dc supply voltage
   d) use a multiband antenna

27. So as to conserve bandwidth, the frequency components of an audio signal being applied to an amplitude modulator should not exceed:
   a) 500Hz
   b) 1kHz
   c) 3kHz
   d) 6kHz

28. The output waveform from an ssb transmission is observed on an oscilloscope. The waveform is "flat topped". This will produce:
   a) 50Hz modulation
   b) splatter
   c) frequency modulation
   d) phase modulation

29. If unwanted rf signals are fed back near a vfo this may cause:
   a) rectification
   b) frequency synthesis
   c) stabilisation
   d) frequency instability

30. Self oscillation of an amplifier is due to:
   a) poor power supply regulation
   b) insufficient amplifier gain
   c) diode switching
   d) coupling between input and output giving a high loop gain

31. A 144MHz transmitter with no harmonic output causes overloading to a nearby uhf tv. This could be overcome by one of the following in the tv antenna lead:
   a) a resistor pad
   b) a low pass filter
   c) a high pass filter
   d) a diode

32. Which of the following might help reduce rf pick up in a loudspeaker lead?
   a) decoupling capacitors in series with the loudspeaker lead
   b) an rf choke wired across the loudspeaker lead
   c) a decoupling capacitor across the loudspeaker leads
   d) a resistor in series with the loudspeaker lead

33. Which of the following filter characteristics would be useful in rejecting an unwanted signal at the input to a receiver?

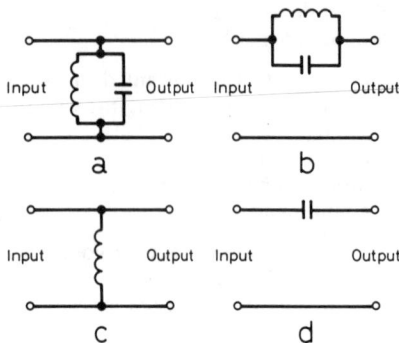

a   b   c   d

34. In order to reduce the risk of mains borne interference, the earth for an end fed antenna should be:
    a) a water pipe in the house
    b) the mains earth
    c) left disconnected
    d) a good separate rf earth

35. A braid breaking choke in a tv coaxial cable down lead will help reduce:
    a) all ac signals
    b) out of phase interfering rf signals
    c) in phase interfering rf signals
    d) mains hum

36. Electromagnetic compatibility is:
    a) two antennas matching each other
    b) two Yagis fitting together mechanically
    c) the ability of equipment to function satisfactorily in its own environment without introducing intolerable electromagnetic disturbances
    d) the disability of equipment to function satisfactorily in its own environment and producing electromagnetic disturbances

37. If two transmissions are received in a non-linear device then:
    a) they will cancel each other out
    b) neither will be received
    c) they must be transformer coupled
    d) inter modulation products are produced

38. On an amateur receiver, unwanted signals are found approximately every 15.625kHz, this is probably due to:
    a) a low frequency government station
    b) unwanted radiation of a tv line time base
    c) wanted radiation of a tv line time base
    d) none of these

39. The position of a braid-breaker in a 15 metre tv down lead should be:
    a) at the tv antenna input
    b) at the antenna
    c) reversed
    d) 1m from the antenna

40. When inserting a high pass filter in a uhf tv down lead, the insertion loss at hf should be:
    a) as low as possible
    b) 75 ohms
    c) as high as possible
    d) ideally zero

41. Equipment is said to be tolerant of rf fields up to 0.5V/m. What should be the maximum erp for a distance of 25 metres?
    a) 3.17W
    b) 10W
    c) 12.5W
    d) 31.7W

42. Audio breakthrough is caused on a neighbour's stereo cassette player when you are using an ssb transmission. This might be minimised by:
    a) changing to a.m
    b) changing to fm
    c) speaking more slowly
    d) speaking faster but longer

43. When living in a densely populated neighbourhood and to keep good relationships with people it is wise to:
    a) always use maximum erp
    b) point the beam at the maximum number of tv antennas
    c) only transmit during the "soaps"
    d) use minimum erp necessary

44. When someone in the neighbourhood complains of tvi it is:
    a) wise to check the log to see if it coincides with your transmissions
    b) wise to deny all responsibility
    c) best to immediately blame the other equipment
    d) mandatory to call the DTI

45. If it is proved that interference is getting into equipment via the mains, it is necessary to:
    a) inform the CEGB
    b) fit a mains filter
    c) swap over the line and neutral connection
    d) disconnect the earth

Sample examination 7, Paper 2
Operating practices, procedures and theory.

1. You are having trouble with reception due to static. The Q code used would be:
    a) QSL
    b) QRX
    c) QRZ
    d) QRN

2. When calling CQ in morse, the transmission should be terminated by:
    a) K

b) KN
c) AR
d) CT

3. It is good practice for:
a) plastic piping to be used as an earth
b) all metal cases to be unearthed
c) there to be no master switch
d) all power to be supplied via a master switch

4. When working through a satellite use:
a) as much power as possible
b) Esperanto
c) sufficient power to maintain reliable communication
d) fm only

5. If a readability report of 1 is given, this would indicate:
a) unreadable
b) readable with considerable difficulty
c) readable with practically no difficulty
d) perfectly readable

6. To obtain an indication if a particular international path is open on hf:
a) call CQ the country on any hf band
b) look at the propagation forecasts
c) telephone the licensing authority
d) examine the weather forecasts

7. Using the International Phonetic Alphabet, TOWN would be:
a) tango, ontario, washington, november
b) tango, oscar, washington, november
c) tango, oscar, whiskey, november
d) tango, ontario, whiskey, november

8. When operating through a repeater:
a) only use cw
b) give priority to mobile stations
c) do not give any breaks between overs
d) give priority to base station nets

9. During telephony transmissions, it is important to:
a) speak clearly and not too quickly
b) always use Q codes
c) always spell names using phonetics
d) speak rapidly

10. The resonant frequency of the above circuit is:
a) 1.59155kHz
b) 15.9155kHz
c) 159.155kHz
d) 1591.55kHz

11. The Q factor in the circuit of Q10 at resonance is:
a) 10
b) 100

c) 1000
d) 10000

12. The peak to peak value of the 240V mains is:
a) 170V
b) 240V
c) 339V
d) 679V

13. The reactance of a choke is given by:
a) $1/2 \pi fL$
b) $2 \pi fL$
c) πfL
d) $2 fL$

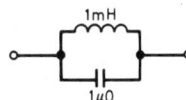

14. The resonant frequency of the above circuit is:
a) 0.5033kHz
b) 5.033kHz
c) 0.5033MHz
d) 5.033MHz

15. In the circuit of Q14, the phase difference between the current in the two branches is:
a) 45 deg
b) 90 deg
c) 180 deg
d) 270 deg

16. Which of the following represents full wave rectification?

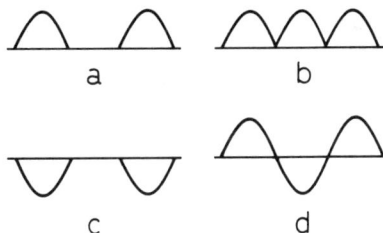

17. The main purpose of a Zener diode is:
a) rectification
b) tuning
c) voltage stabilisation
d) display

18. When a transistor is biased ON, the emitter-base junction is:
a) reverse biased
b) open circuit
c) short circuit
d) forward biased

19. The terminals of a transistor are labelled:

D

a) emitter, base, collector
b) emitter, drain, source
c) drain, source, collector
d) drain, gate, base

20. Which of the following is the correct arrangement for a bridge rectifier?

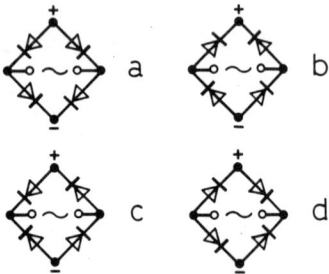

21. In a varactor diode the capacitance:
    a) is constant irrespective of applied reverse voltage
    b) increases as reverse bias is increased
    c) decreases as reverse bias increases
    d) decreases as reverse bias decreases

22. A buffer circuit has which of the following properties?
    a) low input and high output resistance
    b) high input and output resistance
    c) low input and output resistance
    d) high input and low output resistance

23. The main filtering of the wanted signal in a superheterodyne receiver is normally accomplished in the:
    a) rf portion
    b) audio amplifier
    c) mixer
    d) i.f section

24. The advantage of a low i.f is that the filter will have:
    a) high selectivity
    b) broad band characteristics
    c) maximum pass band attenuation
    d) pass band gain

25. For cw reception, the difference in frequency of the bfo and final i.f should be about:
    a) 1kHz
    b) 10kHz
    c) 455kHz
    d) 10.7MHz

26. The squelch in a receiver is operated by:
    a) the vfo
    b) the power supply
    c) the heterodyne oscillator
    d) either the i.f or af signal

27. In some af output stages in a receiver the loudspeaker is driven via a step down transformer. This provides:

a) higher af gain
b) power matching from amplifier to loudspeaker
c) better audio quality
d) economy of the power supply

28. If a station is being received on 14.24MHz and the first i.f of the receiver is 10.7MHz, second channel interference could arise from a signal at:
    a) 10.7MHz
    b) 14.54MHz
    c) 24.31MHz
    d) 35.64MHz

29. In the detection of ssb signals there is normally a carrier insertion oscillator. In a high quality receiver this is:
    a) a vfo
    b) varactor controlled
    c) inductor controlled
    d) crystal controlled

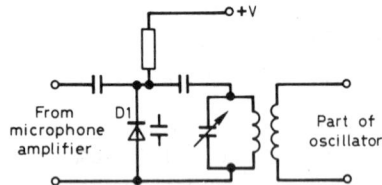

30. The above diagram shows part of an fm transmitter. The purpose of D1 is:
    a) to prevent overloading of the af amplifier
    b) to rectify any rf to provide a dc supply
    c) to vary the oscillator frequency
    d) to provide a dc voltage to drive the relative power output meter of the transmitter

31. When generating an ssb signal, a balanced modulator is used. The filter following this to remove one of the sidebands should have a bandwidth of:
    a) 300Hz
    b) 2.4kHz
    c) 2.4MHz
    d) 10.7MHz

32. The best type of variable capacitor for a vfo would be:
    a) with steel plates
    b) air spaced
    c) as in a cheap transistor radio
    d) with plates separated by polythene

33. Compared to a.m and fm, ssb transmissions are:
    a) of wider bandwidth
    b) easier to demodulate
    c) less adversely affected by disturbances in the ionosphere
    d) easier to produce

34. If the output level of a transmitter is quoted as

20dBW, this is equivalent to:
a) an input power of 20W
b) an output power of 20W
c) an output power of 100W
d) an input power of 100W

35. The efficiency of a power amplifier is the ratio of:
a) rf power out to rf power in
b) rf power input to dc power input
c) rf power output to dc power input
d) rf power input to rf power output

36. In an a.m transmitter using 100% modulation the output voltage can:
a) only be half the average output voltage
b) never exceed the average output voltage
c) rise to twice the average output voltage
d) rise to four times the average output voltage

37. In order to cater for multi-band operation the vfo is normally pre-mixed with another oscillator that is called:
a) a heterodyne oscillator
b) an early oscillator
c) a carrier insertion oscillator
d) a bfo

38. A transmission on 14.18MHz can be received by station A 3000 miles away but not by station B only 40 miles away. This is because:
a) station B is in the skip zone
b) the ground and ionospheric wave cancel out at station B
c) two reflective waves arrive at B with opposite phase
d) propagation conditions are worse than normal

39. Increased ionisation in the D region preventing radio waves from reaching the normal reflecting layers is known as:
a) an aurora
b) sporadic E
c) a critical fade-out
d) a Dellinger fade-out

40. Z_0 for a transmission line is normally called:
a) the velocity factor
b) the characteristic impedance
c) the characteristic factor
d) the copper resistance

41. A wavelength of 10cm in free space corresponds to a frequency of:
a) 3MHz
b) 300MHz
c) 3GHz
d) 30GHz

42. Commonly found values of characteristic impedance for coaxial cables are:
a) 50 and 300 ohm
b) 50 and 75 ohm

c) 75 and 300 ohm
d) 300 and 600 ohm

43. A half wave dipole also resonates at:
a) a sub-harmonic
b) the second harmonic
c) the third harmonic
d) the fourth harmonic

44. As frequency increases, the ionisation to reflect a signal back to earth must:
a) decrease
b) go to zero
c) not change
d) increase

45. The velocity of propagation on a coaxial cable is:
a) greater than in free space
b) the same as free space
c) less than in free space
d) zero

46. Typical dielectric for coaxial cable is:
a) rubber
b) porcelain
c) ceramic
d) polythene

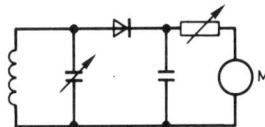

47. The above circuit is representative of:
a) a dip meter
b) an absorption wavemeter
c) a heterodyne wavemeter
d) an oscillator

48. A dummy load for use at vhf should be made from:
a) wire wound resistors
b) carbon resistors
c) metal oxide resistors
d) electric fire heating elements

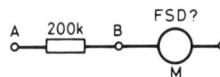

49. For full scale deflection on the above meter with 10V across AB, the FSD should be:
a) $10\mu A$
b) $50\mu A$
c) $100\mu A$
d) $200\mu A$

50. The easiest ac amplitude measurement to take direct from an oscilloscope trace is:
a) rms value
b) average values
c) peak values

d) peak to peak values

51. Which of the following cannot be used to check for harmonics?
    a) a heterodyne wavemeter
    b) an absorption wavemeter
    c) a digital frequency counter
    d) a spectrum analyser

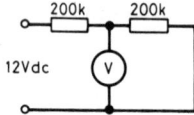

52. In the circuit above the meter is 20 kohm/V and is set to the 10V range. What value will it record?
    a) 0V
    b) 4V
    c) 6V
    d) 8V

53. The typical accuracy of an absorption wavemeter might be:
    a) 0.001%
    b) 0.05%
    c) 1.0%
    d) 5%

54. Which of the following instruments can be used to indicate the matching between a transmitter and a feeder cable?
    a) a multimeter with resistance scale
    b) a reflectometer
    c) a dc resistance bridge
    d) a transformer

55. Which of the following is most suitable to determine the resonant frequency of a trap for a trap dipole?
    a) a frequency meter
    b) a dip meter
    c) an swr bridge
    d) an absorption wavemeter

Sample examination 8, Paper 1
Licensing conditions, transmitter interference and electro-magnetic compatibility.

1. The maximum power permitted in the 432 - 440MHz band is:
   a) 15dBW
   b) 20dBW
   c) 22dBW
   d) 26dBW

2. A nominated member of a user service can operate the station during any disaster relief operation or exercise providing:
   a) they are in the presence of and under the direct supervision of the licensee
   b) they just sign the log book
   c) the other user services agree
   d) the police are in attendance

3. A blank line between entries in a log:
   a) is forbidden
   b) may be left after a CQ call
   c) may be left to indicate bad interference
   d) may be used to insert the call of a second station after contact has been established

4. If a licence has expired, it must:
   a) be returned to the licensing authority
   b) be kept as a memento
   c) be burnt
   d) returned to the police

5. Which of the following need not be put in the log:
   a) date
   b) names of operators worked
   c) frequency band(s)
   d) times of operation

6. The words 'Wireless Telegraphy' used in the licence:
   a) exclude telephony
   b) include telephony
   c) refer only to telephony
   d) refer only to morse

7. The callsign prefix GD should always be used whenever the station is being operated from:
   a) Scotland
   b) Isle of Man
   c) Isle of Dogs
   d) Guernsey

8. The 21MHz band covers:
   a) 21.3 - 21.7MHz
   b) 21 - 21.45MHz
   c) 21 - 21.55MHz
   d) 21.45 - 21.7MHz

9. You are operating a mobile station when you cross from England into Wales. You should:
   a) make a statement that you have done so in the first QSO after the crossing
   b) use the prefix GW only in the first QSO across the border
   c) use and retain the prefix GW for the whole of your time in Wales
   d) do none of these

10. J3E corresponds to:
    a) ssb with full carrier
    b) ssb with no carrier
    c) fm using voice modulation
    d) a cw transmission

11. The amateur radio licence:
    a) gives the right to enter any private property without permission
    b) gives the right to transmit copyright computer programs
    c) does not give the right to enter private property without permission
    d) allows any third party message to be passed

12. The licence states that for frequency control in the sending apparatus:
    a) crystals only must be used
    b) synthesisers must only be used
    c) suitable valves can only be used
    d) satisfactory method of frequency stabilisation shall be employed

13. A station is situated within 0.8km of the boundary of an aerodrome. The antenna system must be arranged so that:
    a) it cannot be seen from the air
    b) it does not exceed 15m above ground level
    c) it is totally within the building housing the station
    d) it does not exceed 15m above the roof of the building housing the station

14. When pronouncing a word phonetically, which of the following is expressly forbidden?
    a) the names of countries
    b) the names of cities
    c) words of an obscene or indecent nature
    d) the names of continents

15. The licence will, without any further payment other than the initial licence fee, be valid for:
    a) ever
    b) one year
    c) five years
    d) ten years

16. To minimise harmonic radiation from a single band uhf transceiver it should ideally be followed by:
    a) a high pass filter
    b) a low pass filter
    c) a band pass filter
    d) a notch filter

17. The output matching circuit and filter in a Class C amplifier should be:
    a) enclosed in a screened box
    b) put on top of the amplifier to aid cooling
    c) encased in perspex so that flashover can be seen
    d) installed at the top of the mast next to the antenna

18. Which of the following filter characteristics would be suitable for following an all band hf transmitter?

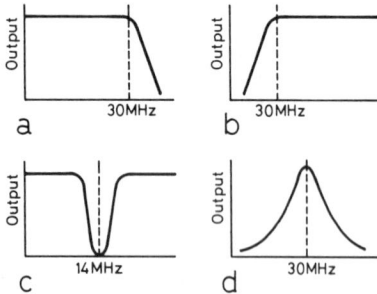

19. Out of band radiation is:
    a) a radiated signal outside the vhf band
    b) a radiated signal outside a designated amateur band
    c) when an electron jumps from one level to the next in an atom
    d) due to too narrow an audio band

20. An absorption wavemeter is useful for:
    a) checking exact transmission frequency
    b) checking frequency drift
    c) checking peak modulation index
    d) checking for harmonic radiation

21. A harmonic is:
    a) a whole number multiple of a frequency
    b) a sub multiple of a frequency
    c) any frequency greater than the original
    d) always an interfering frequency

22. To prevent unwanted radiation from multiplier stages they should be encased in:
    a) metal boxes
    b) epoxy
    c) polythene
    d) wood

23. To reduce the risk of self oscillation in a power amplifier, a screen should be used:
    a) between the output and the mains transformer
    b) between input and output circuitry
    c) between the rectifiers and smoothing capacitors
    d) between all resistors

24. If the frequency stability of a transmitter is poor it may cause:
    a) electric shocks

b) operation out of band
c) excessive collector dissipation
d) excessive power to be drawn from the supply

25. To cut down radiation from a coil in a tuned circuit, it should be:
    a) enclosed in a plastic screening can
    b) enclosed in a non-metallic resin
    c) placed in a metal screening can
    d) mounted external to the equipment

26. An ssb transmitter is likely to cause interference with nearby stations if:
    a) the power amplifier is off
    b) the p.a stage is underdriven
    c) the p.a is followed by a good filter power supply
    d) the power amplifier is overdriven

27. A 'lossy' choke can be used to suppress parasitic oscillations. This would be constructed by:
    a) winding thick copper wire on a very high value resistor
    b) making a self supporting coil of thick copper wire with a silver coating
    c) winding a coil on a low value resistor
    d) using a low value carbon resistor

28. If an L-C oscillator is used to generate directly a signal at 14.05MHz for a cw transmitter and it drifts by −1%, it will:
    a) stay within the designated band
    b) go above the top band edge
    c) go below the bottom band edge
    d) be rejected

29. To prevent unwanted mixing products reaching a transmitter output stage, the output of a mixer should be:
    a) followed by resistive coupling
    b) directly coupled
    c) followed by a transistor
    d) well filtered

30. It is noticed that a Class C amplifier draws a varying current when it is supposedly biased off. This could be due to:
    a) harmonic transmission
    b) critical damping
    c) parasitic oscillations
    d) parasites

31. To minimise the possibility of interference, the power used should be:
    a) the minimum to maintain reliable communication
    b) the maximum permissible
    c) always less than 1W but greater than 500mW
    d) half the maximum permissible

32. A tv receiving on about 570MHz experiences overloading due to a neighbouring 435MHz amateur transmitter. One item that might help is:

a) a low pass filter in the transmitter lead
b) a high pass filter in the transmitter lead
c) a low pass filter in the tv antenna lead
d) a high pass filter in the tv antenna lead

33. Which of the following might be suitable in preventing overloading in an fm broadcast radio from a 70MHz transmission?

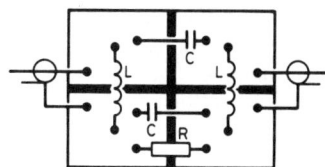

34. When using a pcb drill, interference is experienced on the 14MHz band. The interference the drill generates is classified as:
a) boring
b) swarf-like
c) broad band
d) narrow band

35. One can check for key clicks by using:
a) a receiver tuned to a nearby frequency
b) an absorption wavemeter
c) a digital frequency counter
d) a stethoscope

36. An amateur lives in the middle of a row of terraced houses. For cables within the shack carrying rf it would be wise to:
a) use open wire feeders
b) use screened cables
c) use mains wiring with good pvc covering
d) have resistance wire

37. A 10W transmitter feeds a cable with 0.5dB loss. At the end of this is an antenna with 6.5dB gain. The effective radiated power is:
a) 6W
b) 16.5W
c) 40W
d) 50W

38. An hf transmitter gives 100 watts output to a unity gain antenna when keyed. At a distance of 250 metres this produces a field strength of:
a) 0.28V/m
b) 1.00V/m
c) 28.08V/m
d) infinity

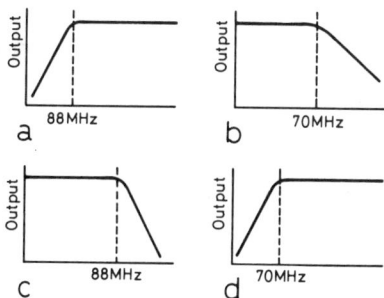

39. The above circuit is typical of:
a) a tv preamplifier
b) a braid breaker with high pass filter
c) an oscillator
d) a braid breaker only

40. In the above circuit R is:
a) to discharge static
b) to provide static
c) for damping oscillations
d) to stop parasitics straying

41. A transformer in a transceiver is fitted with a screen. It should be:
a) left floating
b) connected to Line
c) connected to Line and Neutral
d) connected to a good earth

42. A 50.08MHz transmission causes a large signal at 100.16MHz in the fm broadcast band. To check for emission at this latter frequency one could use:
a) a digital frequency counter
b) an absorption wavemeter
c) an oscilloscope
d) a frequency generator

43. A neighbour complains bitterly that you are causing interference on their new hi-fi. As a first step one could:
a) state that their equipment is at fault
b) invite them in to examine the log book
c) slam the door in their face
d) tell them not to use their equipment when you want to transmit

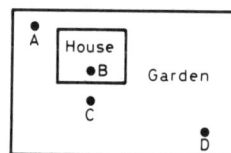

44. In the above plan of house and garden, to minimise the possibility of interference one should site the mast/antenna at:
a) A
b) B
c) C
d) D

45. A "herring bone" pattern on a tv screen is indicative of interference from:

a) a narrow band source
b) a broad band source
c) an automobile
d) a diesel generator

Sample examination 8, Paper 2
Operating practices, procedures and theory.

1. To access a repeater in the UK one must:
   a) send the callsign of the repeater in ASCII
   b) send a 1750Hz tone burst
   c) send an 1850Hz tone burst
   d) speak the callsign of the repeater

2. The abbreviation using the Q code for high power is:
   a) QRH
   b) QRP
   c) QRX
   d) QRO

3. When wearing headphones it is not safe to:
   a) be calling CQ
   b) be switching off
   c) have one's hands inside live equipment
   d) have rubber gloves on

4. An rf choke is commonly connected between the antenna and earth. The purpose of the choke is to:
   a) help tune the system
   b) reject some interference
   c) provide a dc path to prevent high voltage static build up
   d) prevent excessive power being radiated

5. When using voice transmissions, it is wise to:
   a) use jargon continuously
   b) use plain language
   c) do everything in Q codes
   d) speak as fast as possible

6. It is good practice to:
   a) leave pauses between overs on a repeater
   b) leave no pauses between overs on a repeater
   c) let base stations have priority on a repeater
   d) to test a repeater with a tone burst three times before using it

7. In the International Phonetic Alphabet, NAME would be:
   a) nancy, alpha, mike, echo
   b) november, alpha, mexico, echo
   c) november, america, mike, echo
   d) november, alpha, mike, echo

8. Using the RST code, an unreadable signal would be:
   a) T9
   b) R1
   c) R5
   d) S1

9. When operating through a satellite, a station should:
   a) be a member of AMSAT
   b) use the minimum power necessary to maintain communication
   c) always use cw
   d) never call CQ

10. The impedance of this circuit at resonance is:
    a) R
    b) infinity
    c) C
    d) L

11. 0.000000001F is equivalent to:
    a) 1pF
    b) 1nF
    c) $1\mu F$
    d) 1mF

12. The primary of a transformer has five times as many turns as the secondary. If the primary is connected to the 250V mains, what is the expected secondary voltage?
    a) 10V
    b) 25V
    c) 50V
    d) 1250V

13. What is the impedance of the above circuit at 5.033kHz?
    a) zero
    b) 31.62 ohms
    c) 63.24 ohms
    d) 999.8 ohms

14. The phase difference between voltage and current in a purely resistive circuit is:
    a) 0 deg
    b) 45 deg
    c) 90 deg
    d) 180 deg

15. The effective combination of three 33pF in series is:
    a) 11pF
    b) 22pF
    c) 33pF
    d) 99pF

16. The dc current gain of a transistor in the common emitter mode, h_{FE} (or β), is defined by:

a	b	c	d
$\dfrac{\triangle I_E}{\triangle I_C}$	$\dfrac{\triangle I_C}{\triangle I_E}$	$\dfrac{\triangle I_E}{\triangle I_B}$	$\dfrac{\triangle I_C}{\triangle I_B}$

17. A pnp or npn bipolar transistor is:
    a) current controlled
    b) voltage controlled
    c) power controlled
    d) none of these

18. The input resistance of a common emitter amplifier stage is about:
    a) 5 ohms
    b) 50 ohms
    c) 2 kohms
    d) 200 kohms

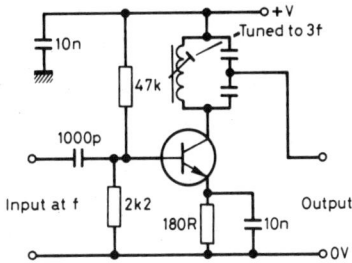

19. The above circuit is that of:
    a) a class C tuned amplifier
    b) an oscillator
    c) a multiplier
    d) a buffer circuit

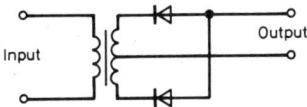

20. The output of the above circuit is:

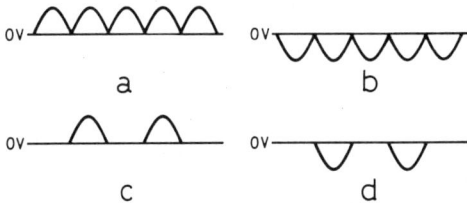

21. To switch a transistor OFF, the base must be:
    a) at the collector potential
    b) at the emitter potential
    c) mid way between collector and emitter
    d) none of these

22. In a semiconductor diode the depletion layer widens with:
    a) decrease in reverse bias voltage
    b) increase in forward bias voltage
    c) increase in forward current flow
    d) increase in reverse bias voltage

23. In order to demodulate cw transmissions in an a.m only receiver, which of the following is required?
    a) a bfo
    b) an fm detector
    c) a crystal multiplier
    d) a morse key

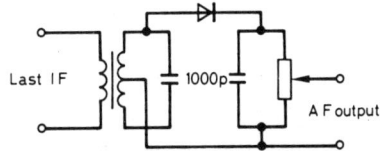

24. The above diagram is that of:
    a) an envelope detector
    b) a mains single phase rectifier
    c) a discriminator
    d) a phase detector

25. To achieve good selectivity when receiving an ssb signal, the bandwidth of the filter should be:
    a) twice that for a.m signals
    b) half that for a.m signals
    c) twice that for fm signals
    d) 1kHz

26. If a receiver uses a final i.f of 455kHz, a suitable bfo frequency to obtain a 1kHz signal when receiving cw is:
    a) 456kHz
    b) 457kHz
    c) 460kHz
    d) 470kHz

27. Which of the following type of bandpass filter is likely to have the narrower bandwidth?
    a) R-C type
    b) L-C type
    c) ceramic resonator
    d) quartz crystal

28. The skirt of a filter is normally specified at about the:
    a) −60dB level
    b) −3dB level
    c) 0dB level
    d) +3dB level

29. Stability of the local oscillator in a transceiver is partially determined by:
    a) good rigid mechanical construction
    b) the use of electrolytics for tuning
    c) the use of resistance wire for the coil
    d) unregulated dc supply to vfo

30. An amplifier is quoted as having a gain of 16dB. The rf output for a 1W rf input should be:
    a) 4W
    b) 16W
    c) 40W
    d) 160W

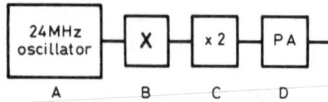

31. The above is a block diagram of a typical 144MHz transmitter using multipliers. The box marked X is:
    a) a doubler
    b) a tripler
    c) a quadrupler
    d) a buffer

32. If the carrier generated by the signal in Q31 is to be amplitude modulated, modulation should be applied at:
    a) A
    b) B
    c) C
    d) D

33. An ideal place to key a cw transmitter is:
    a) the vfo
    b) the whole power supply
    c) the power amplifier
    d) the buffer after the vfo

34. A 145MHz transmitter uses a ×12 multiplier and a deviation of 5kHz. What must the oscillator deviation be to produce this?
    a) 0kHz
    b) 0.417kHz
    c) 5kHz
    d) 60kHz

35. The advantage of J3E emissions is:
    a) power output minimised with modulation
    b) power output maximised with no modulation
    c) minimum power dissipation in the p.a stage with no modulation
    d) maximum power dissipation in the p.a stage with no modulation

36. The coil determining the frequency of a transmitter vfo should be:
    a) air cored with no former
    b) air cored and tightly wound on a former
    c) air cored and loosely wound on a former
    d) wound on a metallic former

37. The above radiation is typical of:
    a) a half wave dipole
    b) a four element beam
    c) a full wave antenna
    d) a three half wavelength antenna

38. A half wave dipole has an impedance of 70 ohms at the centre. It is fed with a half wavelength of 300 ohm ribbon cable. The impedance at the input to the feeder is:
    a) 70 ohms
    b) 185 ohms
    c) 300 ohms
    d) 370 ohms

39. Which of the following represents 100% amplitude modulation?

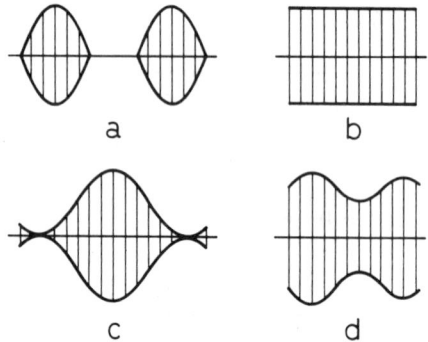

40. A quarter wave transmission line is open circuit at one end, the impedance looking in at the other is:
    a) almost zero
    b) the characteristic impedance
    c) three times the characteristic impedance
    d) infinity

41. The unfolded length of a folded dipole is:
    a) one half wavelength
    b) one wavelength
    c) two wavelengths
    d) four wavelengths

42. A correction factor is normally used in calculating the length of a wire antenna. This is:
    a) 65%
    b) 75%
    c) 80%
    d) 95%

43. The addition of more directors on a beam antenna will:
    a) broaden the beamwidth
    b) raise the radiation resistance
    c) narrow the beamwidth
    d) give a lower front to back ratio

44. The skip zone is where the ground wave:
    a) is enhanced
    b) is reflected
    c) has diminished and the reflected wave has not returned to earth
    d) and reflected wave combine

45. The m.u.f for a given radio path is:
    a) the mean of the maximum and minimum usable

frequencies
b) the maximum usable frequency
c) the minimum usable frequency
d) the mandatory usable frequency

46. A folded dipole has a radiation resistance of about:
   a) 50 ohms
   b) 75 ohms
   c) 300 ohms
   d) 1000 ohms

47. A dummy load is made from twelve 600 ohm wirewound resistors in parallel. It would be suitable for:
   a) radio frequencies up to 30MHz
   b) audio frequencies up to 15kHz
   c) vhf use
   d) uhf use

48. As well as crystal controlled counters, receivers can be checked against:
   a) the 50Hz mains
   b) an L-C oscillator
   c) transmissions such as WWV and MSF
   d) an R-C oscillator

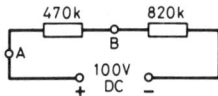

49. To measure the voltage between A and B accurately, which of the following is most suitable?
   a) a moving coil meter of 2 kohms/V
   b) a moving coil meter of 10 kohms/V
   c) a moving iron meter
   d) a digital voltmeter

50. A digital frequency counter has a tolerance of 1 in 10^5. It is used to measure 144MHz transmission. Which of the following has the most suitable number of decimal places for the given accuracy?
   a) 144.375MHz
   b) 144.3752MHz
   c) 144.37522MHz
   d) 144.375221MHz

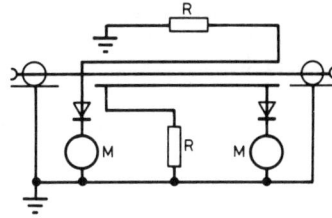

51. The above circuit is that of a simple:
   a) reflectometer
   b) oscilloscope
   c) absolute power meter
   d) absorption wavemeter

52. An oscilloscope can be used to measure:
   a) amplitude
   b) harmonic content accurately
   c) distortion accurately
   d) frequency to 0.001% accuracy

53. An ammeter reads 2A when placed in series with a 50 ohm dummy load. The power in the load is:
   a) 25W
   b) 100W
   c) 200W
   d) 5000W

54. A 12.5V dc power supply is connected to an amplifier. When the rf output from the amplifier is 90 watts the ammeter on the power supply reads 16A. The efficiency of the amplifier is:
   a) 45%
   b) 55%
   c) 100%
   d) 222%

55. The rf output of a transmitter is 100 watts into a matched transmission line. At the antenna end of the line the power is only 50 watts but still at 1:1 vswr. This represents a line attenuation of:
   a) −6dB
   b) 0dB
   c) 3dB
   d) 10dB

Sample examination 9, Paper 1
Licensing conditions, transmitter interference and electro-magnetic compatibility.

1. When conducting a QSO on vhf using F3E, the length of an over should be:
    a) no longer than 2 minutes
    b) no longer than 15 minutes
    c) any length of time without further identification
    d) any length of time with identification at least every 15 minutes

2. The classification R3E is:
    a) ssb with no carrier
    b) ssb with reduced carrier
    c) ssb with full carrier
    d) a.m using double sideband

3. An English station while cycling in Wales should use the callsign:
    a) G7xxx/GW/M
    b) GW7xxx/M
    c) G7xxx/P
    d) GW7xxx/P

4. The station can be closed down at any time by:
    a) a demand from an adjacent amateur
    b) a demand from a neighbour experiencing inter-ference
    c) a demand from a person acting under the authority of the licensing authority
    d) a taxi firm experiencing interference

5. Which of the folowing bands is a class B licensee not permitted use?
    a) 28MHz
    b) 144MHz
    c) 432MHz
    d) 1296MHz

6. Facsimile transmission using a.m is:
    a) A3C
    b) F3C
    c) G3C
    d) A1B

7. Maximum carrier power supplied to the antenna in the 144 - 146MHz band is:
    a) 9dBW
    b) 10dBW
    c) 16dBW
    d) 20dBW

8. An amateur should use the suffix /MM when operating from:
    a) a cruiser on the Norfolk Broads
    b) a barge on the Grand Union Canal
    c) a yacht in the Atlantic
    d) a rowing boat on Loch Ness

9. A class B licence holder has just passed his morse test. To get a class A licence he must:
    a) apply for it anytime in the future
    b) do nothing
    c) apply for it within 12 months or retake the morse test
    d) apply for it within 12 months or retake the RAE

10. The licensee shall not permit unauthorised persons:
    a) to see the equipment
    b) to see the equipment in operation
    c) to repair the equipment
    d) to operate the equipment

11. G1xxx is operating /P from a holiday cottage. He must ensure that:
    a) the log is written in pencil
    b) the location is entered in the log
    c) he leaves blank lines on either side of these entries in the log
    d) the contacts are only entered in the mobile log book

12. The maximum p.e.p permissible on the 1.81 - 2.0MHz band for ssb is:
    a) 15dBW
    b) 20dBW
    c) 22dBW
    d) 26dBW

13. The terms 'messages' and 'signals' mentioned in the Licence:
    a) include visual images by television and facsimile transmission
    b) include visual images by television but not facsimile transmission
    c) include visual images sent by facsimile transmission but not television
    d) do not include visual images sent by television and facsimile transmission

14. In which of the following situations is the use of the station not permitted?
    a) on a public service bus
    b) on an inter-city train
    c) on a private launch on the Thames
    d) in an aircraft

15. Which of the following callsigns is representative of a Northern Ireland class B licensee?
    a) GI2xxx
    b) GI3xxx
    c) GI4xxx
    d) GI6xxx

16. The minimum equipment necessary to check for harmonic radiation is:
    a) a digital frequency counter
    b) an absorption wavemeter
    c) a multimeter
    d) an oscilloscope

17. Parasitic oscillations are caused by:
    a) the effects of stray capacitance/inductance of the wiring
    b) ripple on the power supply
    c) mains-borne interference
    d) poor voltage regulation

18. To prevent rf from a p.a stage feeding back to the vfo the dc supply should be:
    a) af filtered
    b) well filtered at rf
    c) hum free
    d) not filtered at rf

19. To prevent stray radiation from an oscillator it should be enclosed in:
    a) a plastic film
    b) a metal box
    c) a perforated wooden screen
    d) a mica compartment

20. Which of the following represents a band pass filter suitable for suppression of harmonics from a single band transceiver?

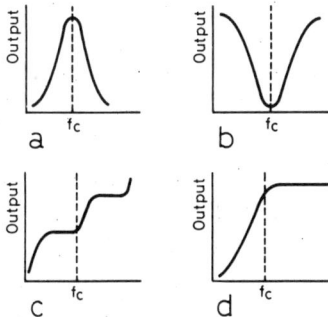

21. The digital readout on a transceiver is only accurate to 0.01%. How close can one go to the lower band edge at 10 metres in order to ensure a carrier only is within band?
    a) 28Hz
    b) 280Hz
    c) 2800Hz
    d) 28000Hz

22. A receiver tuned to which of the following bands will detect a 3rd harmonic from a nearby 7.05MHz transmission?
    a) 14MHz
    b) 21MHz
    c) 24MHz
    d) 28MHz

23. An ammeter in the p.a stage of a transmitter shows slight fluctuations when the transmitter is not being keyed. This possibly indicates:
    a) the presence of parasitic oscillations
    b) good biasing arrangements

    c) the reception of an interfering signal
    d) electromagnetism

24. Interference is experienced in the 144MHz band from some 432MHz crystal controlled equipment. The basic oscillator is around 12MHz. The most likely multiplication is:
    a) ×2×3×3×2
    b) ×2×2×3×3
    c) ×3×3×2×2
    d) ×3×2×3×2

25. A small ferrite bead is sometimes put on a transistor lead. Its purpose is to:
    a) screen the lead
    b) space the transistor above the pcb
    c) prevent parasitic oscillations
    d) give matching to 50 ohms

26. Which of the following pairs of components provide good rf decoupling circuits?
    a) steel cored transformers and ceramic capacitors
    b) ferrite beads and electrolytic capacitors
    c) ferrite beads and polycarbonate capacitors
    d) ferrite beads and ceramic capacitors

27. The minimum equipment for checking the harmonics of a radiated signal is:
    a) a geiger counter
    b) an interval counter
    c) a dip oscillator
    d) an absorption wavemeter

28. The final frequency in a transmitter is obtained by mixing a vfo with a crystal oscillator. Severe drift occurs in the transmission. This is most likely due to:
    a) the crystal oscillator
    b) the vfo
    c) the mixer
    d) the p.a stage

29. In an fm system, if the carrier is over deviated then:
    a) no sidebands are generated
    b) excessive number of sidebands are generated
    c) only two sidebands are generated
    d) only one sideband is generated

30. A frequency synthesiser can generate unwanted radiation unless:
    a) the control circuits are well screened
    b) all the digital circuits are in a plastic enclosure
    c) the output is shorted to ground
    d) none of these

31. To minimise coupling to other antennas in the vicinity, the transmitting antenna should be:
    a) put in the loft space
    b) kept as low as possible
    c) placed underneath the other antennas
    d) kept as far away as possible from the other antennas

32. A strong, but pure, unwanted signal received by a broad band pre-amplifier:
   a) will never cause any problems
   b) may cause blocking
   c) will provide extra gain
   d) will provide extra filtering

33. A public address system is prone to picking up rf signals on its distribution wires. A possible cure is:
   a) putting ceramic capacitors in series with the distribution wires
   b) putting ferrite cored chokes across the distribution wires
   c) putting disk ceramics across and ferrite beads on the distribution wires
   d) using series carbon resistors on the wires

34. When transmitting on 144MHz and 435MHz simultaneously, interference could be caused to a uhf tv due to inherent non-linearities in the tv front end at about:
   a) 291MHz
   b) 579MHz
   c) 650MHz
   d) 1GHz

35. Interference is being caused to a neighbour's hi-fi system. One possible cure would be:
   a) ferrite beads on the transmitter lead
   b) a capacitor across the transmitter lead
   c) screened wire for the loudspeaker leads
   d) open wire feeder for the transmitter lead

36. When living in a densely populated housing estate it might be better not to envisage using:
   a) slow scan tv on 435MHz
   b) fast scan tv on 435MHz
   c) slow scan tv on 1.3GHz
   d) fast scan tv on 10GHz

37. When visiting a neighbour's house to check for tvi, as a first step:
   a) remove the tv mains plug and check the picture
   b) remove the tv antenna lead to see if the interference still persists
   c) switch the tv off
   d) take the back off the tv and earth the chassis

38. When making a mains filter choke:
   a) wind with bare wire
   b) only wind the earth lead on the core
   c) wind the mains lead so that about two thirds of the core is filled
   d) use a polythene core

39. In an electrical system using pme it might be wise to:
   a) disconnect all earths
   b) provide a separate rf earth
   c) cut the neutral connection to the house
   d) connect all rf equipment to the radiators

40. Interference is being caused to a hi-fi receiver. The interference does not vary with rotation of the volume control. Pick up is likely to be:
   a) before the volume control
   b) in the tuner section
   c) in the audio output stages
   d) in the mains lead

41. Interference in the middle of a band is found to be caused by a single carrier. Which of the following could be used to attenuate the interfering signal with minimal degradation to the rest of the band?
   a) a band pass filter
   b) a notch filter
   c) a low pass filter
   d) a high pass filter

42. A radiated signal generates both an electric and a magnetic component. The units of the electric field are in:
   a) metres per volt
   b) volts per ampere
   c) volts per ohm
   d) volts per metre

43. In trying to keep interference to a low level, one is asked to use minimum ERP. ERP stands for:
   a) efficiently radiated power
   b) effective radiated power
   c) electrically radiated power
   d) electrostatic radiated power

44. Insertion loss of a filter is defined as:
   a) reduction of wanted signal in dB when filter is fitted into circuit
   b) reduction of unwanted signal in dB when filter is fitted into circuit
   c) change in impedance in ohms when filter is fitted to circuit
   d) loss sustained by inserting finger into circuit

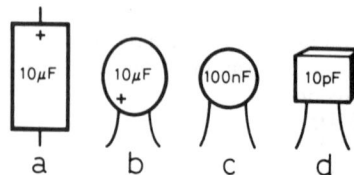

a   b   c   d

45. Which of the above devices is suitable for use as an rf bypass capacitor?
   a) A
   b) B
   c) C
   d) D

Sample examination 9, Paper 2
Operating practices, procedures and theory.

1. Having established contact on a calling frequency it is good practice to:
   a) stay on the same frequency
   b) move to another frequency
   c) invite others to join in on the same frequency
   d) be objectionable to all others calling

2. If a C appears after the report on a cw contact this means:
   a) a crystal controlled signal
   b) chirp on the signal
   c) carrier on the signal
   d) cross-modulation on the signal

3. For safety reasons all exposed metalwork in an amateur station should be:
   a) connected to the live
   b) free of earth
   c) left floating
   d) connected to earth

4. A satellite will:
   a) relay only a single band of frequencies
   b) relay any band of frequencies
   c) need a toneburst for access
   d) only be operated by a secret code

5. CQ calls should only be made:
   a) when band conditions are flat
   b) after listening on a frequency to see if it is not in use
   c) on frequencies that are in use
   d) when there is a contest on

6. If a station asks 'please QSY' this means:
   a) there is fading
   b) change frequency
   c) stop transmitting
   d) reply in morse

7. In the International Phonetic Alphabet, ZEBRA is:
   a) zulu, echo, bravo, romeo, america
   b) zebra, elephant, bravo, romeo, alpha
   c) zulu, echo, bravo, romeo, alpha
   d) zanzibar, echo, bravo, romeo, america

8. Using the RST code, a barely readable signal with only occasional words distinguishable, is:
   a) R1
   b) R2
   c) R3
   d) R4

9. In order to access a UK repeater, it is necessary to:
   a) use a 1750Hz toneburst
   b) use an 1800Hz toneburst
   c) call the repeater using cw
   d) transmit using maximum legal power

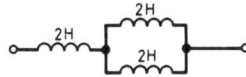

10. Total inductance in the above circuit is:
    a) 1.33H
    b) 3H
    c) 3.5H
    d) 6H

11. The period of a 1MHz wave is:
    a) 1ms
    b) 1μs
    c) 1ns
    d) 1ps

12. The power dissipated in a pure 1μF capacitor at 10kHz with 12V(rms) across it is:
    a) zero
    b) 0.9W
    c) 9W
    d) 90W

13. An inductor of 50μH, a capacitor of 50pF and a resistor of 50 ohms are connected in series. The impedance at resonance is:
    a) 16.66 ohms
    b) 50 ohms
    c) 150 ohms
    d) is frequency dependent

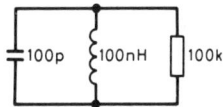

14. The resonant frequency of the above circuit is:
    a) 50.3kHz
    b) 5.03MHz
    c) 50.3MHz
    d) 503MHz

15. A resistor having a value of 100 ohms has 10V across it. Its minimum power rating must be:
    a) 0.125W
    b) 0.25W
    c) 1W
    d) 10W

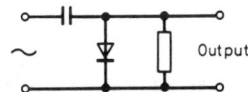

16. An alternating voltage is applied to the circuit shown above. The output waveform will be:

a | b

c | d

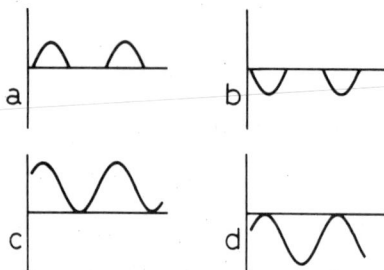

17. A bipolar transistor is:
   a) current controlled
   b) voltage controlled
   c) a thermionic device
   d) a magnetically operated device

18. A class C amplifier operates over:
   a) the complete cycle
   b) three quarters of a cycle
   c) exactly half a cycle
   d) less than half a cycle

19. A typical input resistance for an emitter follower is:
   a) 10 ohm
   b) 1 kohm
   c) 10 kohm
   d) 100 kohm

20. A reverse biased diode exhibits:
   a) no resistance
   b) low resistance
   c) high resistance
   d) high inductance

21. If the base potential of an npn transistor is held at the emitter potential, the collector current will be:
   a) zero
   b) always 1A
   c) between 10mA and 2A
   d) very high

22. P type material has:
   a) an excess of electrons
   b) a deficiency of holes
   c) equal number of holes and electrons
   d) a deficiency of electrons

23. A bfo is used in a receiver to:
   a) make a cw signal using A1A audible
   b) mix with the incoming signal to provide the first i.f
   c) cancel amplitude interference
   d) remove fm signals

24. The reading on an S meter gives an indication of:
   a) vfo voltage
   b) superheterodyne operation
   c) squelch setting
   d) incoming signal strength

25. The advantage of a superheterodyne receiver over a direct conversion type is:
   a) cheaper components can be used
   b) greater selectivity can be achieved
   c) the vfo need not be as stable
   d) it is much simpler in construction

26. In a direct conversion receiver, the local oscillator is:
   a) very much higher than the received signal
   b) very much lower than the received signal
   c) always at 10.7MHz
   d) very close to that of the received signal

27. The frequency difference between the wanted rf signal and the so-called second channel is:
   a) twice the third i.f
   b) twice the wanted rf
   c) twice the first i.f
   d) the wanted rf plus the first i.f

28. In order to minimise variation of audio output with variation of rf input, a receiver is fitted with:
   a) agc
   b) audio gain limiting
   c) audio filters
   d) audio bias control

29. Apart from a discriminator, fm signals can be demodulated by:
   a) a ratio detector
   b) balanced modulators
   c) resistive dividers
   d) Zener diodes

30. If an ssb signal at 9MHz is passed through a frequency multiplier stage it is:
   a) still an ssb signal at a higher frequency
   b) converted to an fm signal
   c) converted to an a.m signal
   d) severely distorted

31. A Class C amplifier following a transceiver can be used to amplify without distortion:
   a) fm, ssb and a.m signals
   b) ssb, a.m and cw signals
   c) ssb only
   d) fm and cw signals only

32. To produce a double sideband, suppressed carrier signal, which of the following should be used?
   a) a balanced modulator
   b) a crystal filter
   c) a single transistor mixer
   d) a single diode mixer

33. In a transmitter vfo it is found that the inductance of the coil rises with temperature and that the capacitor is very stable. The effect of this as temperature rises is:
   a) vfo frequency drifts high
   b) vfo frequency drifts low

c) vfo output remains constant
d) chirp occurs

34. To overcome the problem of Q33, the capacitor should have:
a) a positive temperature coefficient
b) a negative temperature coefficient
c) a high voltage rating
d) a zero temperature coefficient

35. Using audio frequencies up to 3kHz, the bandwidth of a typical a.m transmission with 70% modulation is:
a) 2.1kHz
b) 3kHz
c) 4.2kHz
d) 6kHz

36. To prevent generation of unwanted signals, a mixer should always be followed by:
a) an all stop filter
b) a resistor pad
c) a good filter
d) another mixer immediately

37. When compared with valves, transistors are:
a) more susceptible to overload
b) less susceptible to overload
c) less efficient
d) magnetically controlled

38. The polarisation of an electromagnetic wave is determined by:
a) the position of the transceiver
b) the direction of propagation
c) none of these
d) the orientation of the transmitting antenna

39. What is the length of a piece of coaxial cable cut for a full wavelength at 100MHz if the velocity factor is 0.6?
a) 0.18m
b) 1.8m
c) 3m
d) 18m

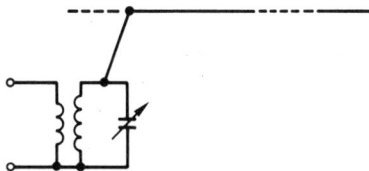

40. The above antenna is representative of:
a) a Yagi
b) a dipole
c) a long wire
d) a monopole

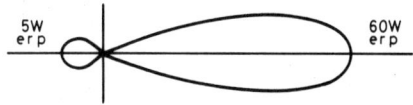

41. The front to back ratio of the antenna depicted by the above radiation pattern is:
a) 10.8dB
b) 12dB
c) 21.6dB
d) 60dB

42. A transformer can be used to match a 75 ohm transmission line to a 300 ohm antenna. The transformer should have a turns ratio of:
a) 1:1
b) 1:2
c) 1:4
d) 1:16

43. A transmission line is said to be perfectly matched when it is terminated by an impedance equal to:
a) half the characteristic impedance
b) twice the characteristic impedance
c) an open circuit
d) the characteristic impedance

44. A balanced transmission line has:
a) one side connected to earth
b) both sides connected to earth
c) a coaxial construction
d) equal impedance of each line to earth

45. When considering the gain of various antennas the term dBi is encountered. This corresponds to dB with respect to:
a) an isotropic radiator
b) a dipole
c) an input
d) an instrument

46. The highest layer in the ionosphere is known as the:
a) D region
b) E layer
c) F1 layer
d) F2 layer

47. Which of the instruments below can measure the exact frequency of a harmonic in a complex waveform?
a) a heterodyne wavemeter
b) a digital frequency counter
c) an absorption wavemeter
d) an oscilloscope

48. Resolution of an instrument is:
a) the smallest division to which a reading can be made
b) how close the instrument is to the true reading
c) the same as the accuracy of the instrument

d) the same as the full scale reading of the instrument

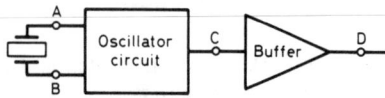

49. When measuring frequency in the above equipment the probe should preferably be placed at:
a) A
b) B
c) C
d) D

50. The moving coil meter relies on:
a) the interaction of a magnetic and an electric field
b) the interaction of two permanent magnetic fields
c) on only one magnetic field
d) the interaction of a permanent magnetic and an electromagnetic field

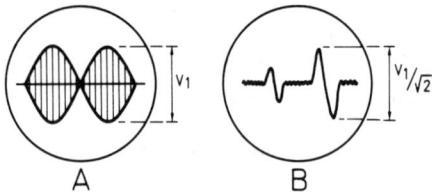

51. On a two tone test on a transmitter the trace at A is produced by a 400W p.e.p signal. What is the value represented by trace B?

a) 100W p.e.p
b) 150W p.e.p
c) 200W p.e.p
d) 280W p.e.p

52. An accuracy of one part in a million is equivalent to:
a) 0.1%
b) 0.01%
c) 0.001%
d) 0.0001%

53. A voltmeter and ammeter are used to measure the dc power to a circuit. The voltmeter reads 10V and the ammeter 1A. If both read low by 5%, the true power taken is:
a) 9.025W
b) 10W
c) 10.5W
d) 11.025W

54. The action of the so called dip oscillator depends on:
a) the extraction of energy from the tuned circuit under test
b) the extraction of energy from the dip oscillator by the circuit under test
c) radiation from a nearby transmitter
d) the tuned circuit under test changing the oscillator frequency of the dip meter

55. Which of the following represents a good "shopping list" when making a dummy load for use at 50MHz?
a) 10 off 500 ohm carbon resistors
b) 1 off 50 ohm wirewound resistor
c) 10 off 500 ohm wirewound resistors
d) 30 turns of heating element wound on a brass former

Answers to questions

Exam	1		2		3		4		5		6		7		8		9	
Part	**1**	**2**	**1**	**2**	**1**	**2**	**1**	**2**	**1**	**2**	**1**	**2**	**1**	**2**	**1**	**2**	**1**	**2**
1	B	C	A	D	A	D	C	B	B	A	A	A	D	D	D	B	D	B
2	D	B	D	D	A	D	D	B	C	D	B	A	A	A	A	D	B	B
3	D	C	D	B	C	D	B	B	B	A	B	A	A	D	A	C	B	D
4	C	B	B	A	B	C	A	A	D	B	B	C	B	C	A	C	C	A
5	D	B	D	A	B	B	B	C	C	C	B	B	A	A	A	B	C	B
6	D	D	B	B	A	D	B	B	D	A	C	D	C	B	B	A	A	B
7	A	C	C	D	D	A	B	B	D	A	A	C	D	C	B	D	D	C
8	B	A	C	B	D	C	A	D	B	C	D	A	C	B	B	B	C	B
9	C	B	D	C	D	B	C	C	C	A	D	D	A	A	C	B	A	A
10	A	A	A	D	C	D	A	B	A	A	C	B	C	C	B	A	D	B
11	C	C	A	C	A	A	A	D	A	A	D	D	B	A	C	B	B	B
12	D	C	B	A	A	C	A	B	B	C	B	C	A	D	D	C	A	A
13	C	B	C	B	C	C	D	D	D	C	D	B	B	B	B	A	A	B
14	B	B	D	B	D	C	B	C	D	D	C	B	B	B	C	A	D	C
15	C	D	D	B	D	D	A	A	C	A	D	C	D	C	B	A	D	C
16	A	D	B	D	B	D	B	C	A	D	B	B	B	B	C	D	B	D
17	B	B	B	A	A	A	A	A	B	D	C	C	A	C	A	A	A	A
18	D	B	A	B	D	A	C	D	C	D	D	C	C	D	A	C	B	D
19	C	D	C	A	D	C	D	B	B	C	D	C	B	A	B	C	B	C
20	B	A	C	C	C	A	A	A	A	C	A	D	C	B	D	B	A	C
21	A	B	B	A	C	B	A	C	A	B	A	B	A	C	A	B	C	A
22	D	B	A	C	A	A	B	C	B	A	D	C	D	D	A	D	B	D
23	B	D	B	D	C	D	B	A	A	A	A	D	A	D	B	A	A	A
24	A	A	D	C	B	D	A	C	A	B	A	C	C	A	B	A	B	D
25	D	B	A	C	A	B	B	B	B	D	C	D	A	A	C	B	C	B
26	B	C	C	A	D	D	C	C	B	C	C	C	B	D	D	A	D	D
27	D	D	C	B	B	B	B	C	A	C	D	A	C	B	C	D	D	C
28	D	C	C	C	B	D	D	B	A	A	B	A	B	D	C	A	B	A
29	C	B	B	C	B	C	B	D	D	A	C	A	D	D	D	A	B	A
30	A	B	C	B	B	B	C	A	C	A	A	B	D	C	C	C	A	D
31	C	C	C	B	C	A	A	C	A	C	C	A	C	B	A	B	D	D
32	D	A	C	D	C	C	C	C	D	B	A	D	C	B	D	D	B	A
33	A	B	C	A	C	D	B	A	D	A	A	A	B	C	A	D	C	B
34	A	C	B	D	A	D	C	A	C	B	A	A	D	C	C	B	B	B
35	B	D	D	D	D	D	D	D	A	B	A	A	C	C	A	C	C	D
36	B	D	A	A	C	D	C	A	B	A	C	B	C	C	B	B	B	C
37	C	B	D	A	B	B	C	B	D	B	B	D	D	A	C	A	B	A
38	B	D	A	A	D	C	C	B	A	D	C	B	B	A	A	C	C	D
39	C	C	C	A	C	B	C	C	A	D	A	A	A	D	B	C	B	B
40	C	C	D	D	A	A	A	D	B	A	B	A	A	B	A	A	C	C
41	D	C	B	A	B	A	B	B	A	B	C	C	A	C	D	B	B	A
42	C	C	D	C	C	D	C	D	B	C	A	C	B	B	B	D	D	B
43	A	C	C	B	A	B	A	C	D	B	A	D	D	C	B	C	B	D
44	A	A	C	C	D	B	D	D	C	A	C	C	A	D	D	A	A	D
45	C	D	C	C	B	C	B	A	A	B	C	C	B	C	A	B	C	A
46		C		A		D		A		B		C		C		C		A
47		C		C		C		D		A		D		B		B		A
48		C		A		D		D		A		C		B		C		A
49		C		B		D		C		C		C		B		D		D
50		C		D		D		B		A		B		D		A		D
51		C		D		A		D		C		D		C		A		C
52		B		C		D		A		C		D		B		A		D
53		B		B		D		D		C		B		A		C		B
54		B		C		C		C		A		C		B		A		B
55				D		D		A		C		B		B		C		A

Recommended reading

The following publications will be of value to those preparing for the RAE.

Radio Amateurs' Examination Manual by G. L. Benbow, G3HB - RSGB.

How To Become A Radio Amateur (BR79) may be obtained free from: Radio Amateur Licensing Unit, Post Office Counters Ltd, Chetwynd House, Chesterfield, Derbyshire, S49 1PF. Tel: 0246-217555/217699. The current edition of this is essential as it contains the latest licence conditions.

How to Improve Television And Radio Reception may be obtained free from Post Offices.

Please contact the RSGB Membership Services Department, Lambda House, Cranborne Road, Potters Bar, Herts, EN6 3JE, for further details of RSGB membership, its publications and its journal *Radio Communication*.

Some other RSGB publications...

PRACTICAL WIRE ANTENNAS
Wire antennas offer one of the most cost-effective ways to put out a good signal on the HF bands, and this practical guide to their construction has something to interest every amateur on a budget. Theory has been kept to a minimum – instead, the author has shared his years of experience in this field.

HF ANTENNAS FOR ALL LOCATIONS
This book explains the ''why'' as well as ''how'' of hf antennas, and takes a critical look at existing designs in the light of latest developments.

AMATEUR RADIO AWARDS (third edition)
This new edition of Amateur Radio Awards gives details of major radio amateur awards throughout the world. Each award is listed in an easy to understand format giving all the information on how to achieve the award. An innovation for this edition is the provision of checklists so that the amateur can keep a record of progress. This book is essential reading for the avid award hunter and the dx chaser alike.

AMATEUR RADIO OPERATING MANUAL
Covers the essential operating techniques required for most aspects of amateur radio including station organisation, and features a comprehensive set of operating aids.

RADIO COMMUNICATION HANDBOOK
First published in 1938 and a favourite ever since, this large and comprehensive guide to the theory and practice of amateur radio takes the reader from first principles right through to such specialised fields as radio teleprinters, slow-scan television and amateur satellite communication.

WORLD PREFIX MAP
This is a superb multi-coloured wall map measuring approximately 1200mm by 830mm. It shows amateur radio country prefixes worldwide, world time zones, IARU locator grid squares, and much more. A must for the shack wall of every radio amateur.

RADIO SOCIETY OF GREAT BRITAIN
Lambda House, Cranborne, Road, Potters Bar, Herts. EN6 3JE

RSGB — Representing Amateur Radio . . .

Radio Communication

A magazine which covers a wide range of interests and which features the best and latest amateur radio news in its special Bulletin section. The Society's journal has acquired a world-wide reputation for its content. It strives to maintain its reputation as the best available and is now circulated, free of charge, to members in over 150 countries. The regular columns in the magazine cater for hf, vhf/uhf, microwave, swl, clubs, satellite, data, contests, and amateur TV. In addition to technical articles, the highly regarded Technical Topics feature caters for those wishing to keep themselves briefed on recent developments in technical matters. Major issues are discussed each month in the editorial column.

The "Last Word" is a lively feature in which members can put forward their views and opinions and be sure of receiving a wide audience. To keep members in touch with what's going on in the hobby, events diaries are published each month.

Members' Advertisements

Subsidized advertisements for the equipment you wish to sell in the Society's monthly magazine, with short deadlines and large circulation.

QSL Bureau

Members enjoy the use of the QSL Bureau free of charge for both outgoing and incoming cards. A leaflet is available from HQ.

Special Event Callsigns in the GB series are handled by RSGB. They give amateurs special facilities for displaying amateur radio to the general public. For details an application form, apply to the Membership Services Department at HQ. Please apply at least six weeks in advance.

Specialized News Sheets

The weekly DX News-sheet for HF enthusiasts, the VHF/UHF Newsletter for VHF enthusiasts, the Microwave Newsletter for those operating above 1GHz and Connect International for packet radio enthusiasts. Details on request from the Circulation Department at HQ.

Specialized Equipment Insurance

Insurance for your valuable equipment which has been arranged specially for members. The rates are very advantageous: details from HQ.

Audio Visual Library

Films, audio and video tapes are available through one of the Society's Honorary Officers for all affiliated groups and clubs. Further details may be obtained either from the Honorary Officer (whose name can be found in Radio Communication) or from the Membership Services Department at HQ.

Reciprocal Licensing Information

Always contact the Membership Services Department at HQ as early as you can if you plan to go abroad. Details are available for most countries on the RSGB computer data base.

Government Liaison

One of the most vital features of the work of the RSGB is the ongoing liaison with the UK Licensing Authority - presently the Radiocommunications Division of the Department of Trade and Industry. Setting and maintaining the proper framework in which amateur radio can thrive and develop is essential to the well-being of amateur radio. The Society spares no effort in defence of amateur radio's most precious assets - the amateur bands.

Beacons and Repeaters

The RSGB supports financially all repeaters and beacons which are looked after by the appropriate committee of the Society, ie, 1.8-30MHz by the HF Committee, 30-1000MHz (1GHz) by the VHF Committee and frequencies above 1GHz by the Microwave Committee. For repeaters, the Society's Repeater Management Group has played a major role. Society books such as the Amateur Radio operating Manual give further details, and computer based lists giving operational status can be obtained by post from HQ - see the price list for details of how to obtain these.

. . . Representing You !

Operating Awards

A wide range of operating awards are available via the responsible officers: their names can be found in the front pages of Radio Communication and in the Society's Members Handbook. The RSGB also publishes a book which gives details of most major awards.

Contests (HF/VHF/Microwave)

The Society has two contest committees which carry out all work associated with the running of contests. The HF Contests Committee deals with contests below 30MHz, whilst events on frequencies above 30MHz are dealt with by the VHF Contests Committee.

Morse Testing

In April 1986 the Society took over responsibility for morse testing of radio amateurs in the UK. If you wish to take a morse test write direct to RSGB HQ (Morse tests - BR) for an application form.

Slow Morse

Many volunteers all over the country give up their time to send slow morse over the air to those who are preparing for the 12 words per minute morse test. You can find the schedule in Radio Communication, or the Members Handbook. The Society also produces morse instruction tapes.

RSGB Books

The Society publishes a range of books for the radio amateur and imports many others. The price list and ordering details can usually be found at the back of Radio Communication. RSGB members are entitled to a 15% discount on all books purchased from the Society. This discount can offset the cost of membership.

Propagation

The Society's Propagation Studies Committee is highly respected - both within the amateur community and professionally - for its work. Predictions are given in the weekly GB2RS news bulletins, the Society's monthly magazine Radio Communication.

Technical Advice

Although the role of the Society's Technical and Publications Committee is largely to vet material intended for publication, its members and HQ staff are always willing to help with any technical matters.

EMC Advice

Breakthrough in domestic entertainment equipment can be a difficult problem to solve as well as having licensing implications. The Society's EMC Committee is able to offer practical assistance in many cases. The Society also publishes a special book to assist you. Additional advice can be obtained from the EMC Committee Chairman via RSGB HQ.

Planning Permission

There is a special booklet and expert help available to members seeking assistance with planning matters.

GB2RS

A special radio news bulletin transmitted each week and aimed especially at the UK radio amateur and short wave listener. The script is prepared each week by the Society's HQ staff, and items of news can be left on the special telephone answering machine on 0707 59260. The transmission schedule for GB2RS is printed regularly in Radio Communication, or it can be obtained via the Membership Services Department at HQ. It also appears in the Members Handbook. The GB2RS bulletin is also sent out over the packet radio network.

Raynet (Radio Amateur Emergency Network)

Several thousand radio amateurs give up their free time to help with local, national and sometimes international emergencies. There is also ample opportunity to practice communication and liaison skills at non-emergency events, such as county shows and charity walks, as a service to the people. For more information or details of how to join, contact the Membership Services Department at HQ.

Notes